AF776457

THE SOULS OF YORUBA FOLK

BLACK STUDIES & critical thinking

SPIRITUALITY & INDIGENOUS THOUGHT

Cynthia Dillard
Series Editor

Rochelle Brock and Richard Greggory Johnson III
Executive Editors

Vol. 70

The Black Studies and Critical Thinking series
is part of the Peter Lang Education list.
Every volume is peer reviewed and meets
the highest quality standards for content and production.

PETER LANG
New York • Bern • Frankfurt • Berlin
Brussels • Vienna • Oxford • Warsaw

TEMITOPE E. ADEFARAKAN

THE SOULS OF YORUBA FOLK

Indigeneity, Race, and Critical Spiritual Literacy in the African Diaspora

PETER LANG
New York • Bern • Frankfurt • Berlin
Brussels • Vienna • Oxford • Warsaw

Library of Congress Cataloging-in-Publication Data

Adefarakan, Temitope E., author.
The souls of Yoruba folk: indigeneity, race, and critical spiritual literacy
in the African diaspora / Temitope E. Adefarakan.
pages cm. – (Black studies and critical thinking; v. 70)
Based on the author's thesis (doctoral)–University of Toronto, 2011, issued under the title:
Yoruba indigenous knowledges in the African diaspora:
knowledge, power and the politics of indigenous spirituality.
Includes bibliographical references and index.
1. Yoruba (African people)–Canada–Ethnic identity. 2. Yoruba (African people)–Religion.
3. Yoruba diaspora. 4. African diaspora–Psychological aspects. 5. Ifa (Religion)–Canada.
6. Cosmology, African. I. Title. II. Series: Black studies & critical thinking; v. 70.
F1035.Y67.A24 305.896'333071–dc23 2015008431
ISBN 978-1-4331-2609-3 (hardcover)
ISBN 978-1-4331-2608-6 (paperback)
ISBN 978-1-4539-1583-7 (e-book)
ISSN 1947-5985

Bibliographic information published by **Die Deutsche Nationalbibliothek**.
Die Deutsche Nationalbibliothek lists this publication in the "Deutsche
Nationalbibliografie"; detailed bibliographic data are available
on the Internet at http://dnb.d-nb.de/.

© 2015 Peter Lang Publishing, Inc., New York
29 Broadway, 18th floor, New York, NY 10006
www.peterlang.com

All rights reserved.
Reprint or reproduction, even partially, in all forms such as microfilm,
xerography, microfiche, microcard, and offset strictly prohibited.

Fún Bàbátúndé
Ki Oluwá ko fí ifẹ Rệ sí ọkan rệ. Kí O si mú ilerí rệ ṣệ nínú aiyé rệ
AṢẸ

For Babatunde
May you rise to your Destiny and always love your Whole Self.
AṢẸ

CONTENTS

SERIES EDITOR'S PREFACE

"A child doesn't belong to the father or the mother: A child belongs to the ancestors." This ancient African proverb bears cosmological witness and offers a scholarly imperative to (re)member our existence as Black people and the central work of providing education that engages our histories, our present conditions, and our futures. But this call is also a recursive one: It is a call to look inward, to look backward, and to look forward for understandings of *how* to move in ways pleasing to those on whose shoulders we stand. *The Souls of Yoruba Folk: Indigeneity, Race, and Critical Spiritual Literacy in the African Diaspora* brings together all of these dimensions in an important example of what Adefarakan describes as a "personal and political effort to de-pathologize African spirituality and shift the popular imagination (African and otherwise) from its profound reliance on re-circulated colonizing scripts about African Indigenous spirituality, to ones that are affirming, nuanced, and grounded in an African-centered, feminist, and anti-colonial politic." Locating this work in the conceptual tradition of W. E. B. DuBois' notion of double-consciousness, this text raises and inquires into the important, often contradictory, and always complicated ideals that DuBois raised for the African in America, namely the unreconciled (and possibly unreconcilable) nature of the souls of Black people who live in diasporic spaces, spaces that too often render us inferior through dominance

and oppression of our minds, bodies, and souls. Here, Adefarakan adds to the spiritual and scholarly legacy set by DuBois, pointing to the ways that colonization within Western thought have produced scant studies in social sciences, education, and cultural studies by or about African spiritualities in diaspora. More specifically, as one of the most popular expressions of Indigenous spirituality in the African diaspora, Adefarakan shines a light on the soul of Yoruba folk as a nuanced "diasporic indigeneity." In important ways, she extends DuBois' notions and our understandings of double consciousness and spiritual strivings to ones that are not solely bound by physical space or location, but as realities in diasporas as well. I was particularly struck by ways that the Yoruba elders and community member participants in this text made sense of being Yoruba people who often overtly pushed down or masked Yoruba spirituality, religion, and traditions while simultaneously lifting up Christianity and/or Islam as their "only" forms of religious practice in Canada. What we learn in *The Souls of Yoruba Folk: Indigeneity, Race, and Critical Spiritual Literacy in the African Diaspora* helps us to better understand these seemingly contradictory realities and points us to the serious consequences and costs of being Black in diaspora(s), where forgetting is often encouraged, "necessary," and a political strategy of survival.

Each chapter of *The Souls of Yoruba Folk: Indigeneity, Race, and Critical Spiritual Literacy in the African Diaspora* is one that causes you to pause as Adefarakan deftly turns our taken-for-granted and often narrow lenses of spirituality and helps us to open them wider, especially in her study of those often lumped into the collective "African," or "American," or "indigenous" and extends it to include the "Nigerian," the "Canadian," and the "Yoruba." In doing so, we come to see the powerful nature of what it means to be spiritual in the African diaspora a little bit more clearly and expansively. In the final chapter of this critical work, Adefarakan pens a letter to teachers, putting forth moral and ethical dilemmas and imperatives for our work of teaching children with Indigenous African spiritualities in mind and heart. With passion grounded in her research as well as her personal identity as a Yoruba woman teacher and mother, she argues the need for children in general and Black children more specifically to have deep spaces within the school context to develop skills of critical spiritual literacy. As I have argued in my own work, such spaces and skills, by necessity, must be centered on helping students "learn to (re)member the things they've learned to forget" in order to make *informed* decisions about the direction and desires within their spiritual identities and lives.

I want to say a few words about the cover images found on *The Souls of Yoruba Folk: Indigeneity, Race, and Critical Spiritual Literacy in the African Diaspora*. Adefarakan selected the beautiful *gelede* of the Ketu-Yoruba people. The gelede performance is an important and sacred ceremony that pays homage to the spiritual powers of women, especially the elder women known by the Yoruba as "our mothers." Through performances of music, dance, song, and the actual mask and dress, the gelede maskers appeal to and encourage the community and our mothers to use their extraordinary powers for the well-being of the society. Analytically, this performance can be seen as a form of commentary on social matters of the community that is at once secular and sacred. And this powerful book that the reader holds in their hands is an echo of the same spirit. Adefarakan helps us to understand the nuances of indigeneity, race, and spirituality amongst the Yoruba in Canada. In a profoundly endarkened feminist way, she also encourages us to (re)cognize the gifts of our mothers and to (re)member that they are, as the Yoruba believe, the "owners of the world." Thus, we must (re)member not for ourselves: We must do this for our children. *The Souls of Yoruba Folk: Indigeneity, Race, and Critical Spiritual Literacy in the African Diaspora* is a brave text. May we listen carefully to the wisdom of the ancestors as they teach us through this important volume.

Cynthia B. Dillard, Ph.D. (Nana Mansa II of Mpeasem, Ghana, West Africa)
Series Editor, Peter Lang, Black Studies and Critical Thinking Series—Spirituality and Indigenous Thought

ACKNOWLEDGMENTS

Modupe Olorun Eleda mi [I give thanks to God my Creator].

To the Most High *Olorun Olodumare*, the Divine Supreme Being and Creator of our Universe, I am grateful for the strength and courage endured through the years it took to journey *The Souls of Yoruba Folk*, and especially for the gift of my son along the way. I am grateful for the many people sent during this time who joined me on the journey, at different points, for different reasons and purposes. Some are still with me, while others have diverged to take their own paths. I give thanks to have experienced their presence in my life and the imprints they have left both in my life and this book.

I thank the 16 Yoruba community members for their indispensable contribution, who, upon agreeing to participate in this project, made it rich and unique. This book could not have happened otherwise: Thank you for having the courage to voice your experiences and for sharing them with me. I am also grateful to Toronto's Yoruba Community Association (YCA), whose staff and members have been extremely supportive.

What is now *The Souls of Yoruba Folk: Indigeneity, Race, and Critical Spiritual Literacy in the African Diaspora* began as my doctoral dissertation; hence, I remain grateful to my thesis committee: Dr. George Dei, my thesis supervisor and mentor, I thank you for the staunch critical support and unwavering belief in the importance of this topic. Dr. Tara Goldstein, I am eternally grateful for your scholarly expertise and overall support primarily because your door was always open. Never did you doubt that I could do this and your kind and encouraging words always kept me going. Dr. Martin Cannon, thank you for your support and keen ability to critically engage my work in such a way that it strengthened my scholarship and challenged me to think about research from various perspectives, especially from that of the Indigenous peoples of Turtle Island.

Dr. Cynthia B. Dillard … where do I begin? A woman who initially was the external examiner for my doctoral dissertation, I am ever so grateful for your sistership and mentorship in this work, *as lived praxis*. Simply put, this book would not have been possible without your patient support and confident faith that it would happen. And it did. *Mo de dupe* [So I thank you]. Your scholarship has been indispensable as an exemplar of writing about spiritual life in the academy, and I continue to learn and grow, taking cues from your life as an African feminist academic grounded in our Indigenous spirituality.

To my family: My mother who gave me life and recently saved it—literally and figuratively—I have come full circle and am a better woman, mother, and person for it. Thank you does not begin to quantify what you have done for me but I can say this: Every day I aspire to be even half the courageous, strong, loving, caring, and selfless woman, mother, and human being you are. Thank you for having faith in me, and having it *for* me when I had none for myself. Thank you also to my father, who was there and has supported my work. To my dear big brother Adedayo, words do not begin to explain the significance you hold in my life so I will say this: I pray every day that my son grows into a man with a spirit, heart, and soul like yours. If, in raising him, I can accomplish this, then I know my son will be the confident, loving, and caring Black man he is supposed to be so that he can live a happy and fulfilling life. Thank you for being who you are. To my sister Adeola, thank you for challenging and teaching me in the unique way that you do. I continue to wish only the best for you and Jumoke. LJC, you always said you came into my life for a purpose, so I thank you for being the vehicle through which Babatunde was blessed into my life.

To Babatunde, my miracle child: "*Okunrin Meta, okunrin Ogun!*" You are my greatest joy and every day I am amazed by you. Your presence in my life has forever enriched it. A greater gift I could never have received. So I thank you for always putting a smile on my face and for coming into my life at the right moment in time. I look forward to raising you and watching you grow.

To those who provided me with emotional support through the many twists and turns that life unfolded before me: all my sister friends and especially to Esther Addo, Sheila Batacharya, Nneka MacGregor, and Gail and Kizzy Bedeau, I am grateful for your presence in my life and cherish your friendship.

A special thank you to the small but ever-so-powerful community of women who are proudly grounded in our African spiritual traditions, Roberta Timothy, Marilyn Johncilla, and Mercedes Umana, thank you for being sisters in the struggle and for the conversations, advice, support, and friendship. May *Olodumare*, our *Orisa*, and our Ancestors continue to guide, protect, and bless us and our children and may we continue to open ourselves to their blessings. Aṣẹ.

Awon egungun mi, ati awon Iya ati Baba mi [To my Ancestors, Mothers and Fathers] of generations most recent to generations most past, you who cleared the path and sacrificed for me, I am forever in your debt and will not forget. So I pledge this book, *The Souls of Yoruba Folk*, as a small offering to that indebtedness. I hope I have done you justice in this call that your grand souls be remembered. Thank you for providing all that you have and lovingly nudging me to the "finish" line.

They say it takes a village to raise a child; it also takes a village to write a book. Thank you to the village—seen and unseen—who congregated in the culmination of this work. It was a communal labor of love. AṢẸ

Permissions

One chapter in this book has previously appeared in the following publication:

Chapter 2 is a slightly revised version of Adefarakan, T. (2011). (Re)Conceptualizing "Indigenous" from Anti-Colonial and Black Feminist Theoretical Perspectives: Living and Imagining Indigeneity Differently. In Dei, G. J. S. (Ed.). *Indigenous Philosophies and Critical Education: A Reader*. New York: Peter Lang. The author gratefully acknowledges Peter Lang Publishing Inc. for its permission to use this chapter.

· 1 ·

A CALL ... TO THE SOULS OF YORUBA FOLK

Many stories of the Yoruba have been told: stories by artists, often in the form of written literature, visual art, poetry, and theater; stories on film, namely from Nollywood; stories by academics, historically from an anthropological perspective that framed us Yoruba within the grand old metanarrative of Africans as primitive, subhuman, and uncivilized objects of study in need of Western (read: White) aid. More recently, academic stories of Yoruba peoples are being told by Africans (who are also often Yoruba) in an attempt to rescue ourselves from those Eurocentric narratives that have characterized us as bankrupt of our humanity. Many of these stories have been told with the intent to write back to empire, humanizing us while simultaneously telling of the atrocities of transatlantic slavery, colonialism, racism, and sexism. These stories proliferate and continue to grow, and rightly so. Woven in and out of these artistic and academic narratives have been those told by the griots, the Babalawos and Iyalawos, the storytellers sitting by the fireside telling Yoruba trickster tales and fables to the young ones as they listen intently. Some narratives have been captured, reimagined, and creatively re-spun by artists and academics alike, while others have likely been lost to us and died with our wisest sages. Inevitably, there are many stories of the Yoruba. And what I offer here in *The Souls of Yoruba Folk: Indigeneity, Race, and Critical Spiritual Literacy in the African Diaspora* is another story of this dynamic African people.

Hinged on the beckoning urgency of articulating spiritual life as experienced in the racialized Black body, this book tells the story of a contradiction: It explores how Indigenous Yoruba spirituality (otherwise known as Ifa) can arguably be held as the most popular expression of African spirituality in its diaspora (M. Asante, personal communication, 2012), yet it simultaneously can also be held as a form of spirituality that most Yoruba speaking peoples themselves do not embrace (at least overtly so). Instead, many Yoruba profess to and proclaim a Christian or Muslim identity as their only form of religious observance. Save a handful of Yoruba elders, some of which are Wole Soyinka (2012, 1981, 1959), Wande Abimbola (1997a, 1997b), and scholars such as Jacob Olupona (2000), Afrocentric scholar Molefi Asante, and African/Black feminist Bibi Bakare-Yusuf (2003), this contradiction has rarely been discussed and given little extensive attention, academic or otherwise. What is compelling is to ask, Why is this so? This book offers a response in the form of a nuanced reading and articulation of this contradiction through a closer examination of Yoruba migrants' spiritual lives in the African diaspora. It does this through the term *critical spiritual literacy* and uses this concept to investigate what we currently know about Yoruba spiritual life, especially as it is lived in the racialized Black body.

In their highly influential book, *Spiritual Literacy: Reading the Sacred in Everyday Life*, Frederic and Mary Ann Brussat (1996) define spiritual literacy as "the ability to read the signs written in the texts of our own experiences … and find sacred meaning in all aspects of life" (15). They identify children and Indigenous peoples as some of the most spiritually literate people in our world despite the fact that "they may not be able to read letters on a page" (18). What the Brussats point out is the significance of different forms of literacy, particularly where spirituality is concerned. In other words, this form of literacy extends beyond conventional Eurocentric notions of being able to read, which are overwhelmingly confined to comprehension of letters on a page.

Crucial to note is that the Brussats' (1996) discussion of various blocks to spiritual literacy is based on their understanding that the sacred is not separate from daily life. Brussat and Brussat identify worldview as the most significant block to spiritual literacy, and while they do not name or identify this blockage as Christian worldview, they do discuss some of the problems that arise from imbibing a perspective where "the world is seen as Devil-ridden, doomed, and dangerous" (33). They argue that embracing such a worldview blocks one's ability to see, commune with, and engage the sacred in daily life. This perspective is strikingly similar to my discussion throughout this book, where I point out that such beliefs about the world are overwhelmingly present in Christian doctrine and worldview. Ultimately, the point here is that worldview or cosmology is of utmost importance where spiritual literacy and matters of the soul are concerned. Through their conceptualization of spiritual literacy the Brussats have made an important contribution to how we think about this in itself. However, their understanding leaves too many gaps that cannot adequately address inequities that arise from oppression.

Consequently, my conception of *critical* spiritual literacy builds on the Brussats' to consider the historical and contemporary power inequities that challenge the learning and use of Yoruba Indigenous knowledges by diasporic Yoruba peoples and their communities. These inequities are largely colonial; hence, I place extreme importance on Indigenous peoples' abilities in terms of their freedom to commune with and recognize the sacred in daily life. Therefore, critical spiritual literacy involves an appreciation of the deep connections between language, cosmology, and worldsense (Oyewumi, 1997) and one's collective and individual identities. Accordingly, this book explores how the sacred is embedded within, and written into, our everyday experiences and lives. In particular, *The Souls of Yoruba Folk* examines how Yoruba migrants in the African diaspora engage with and read the sacred in their daily lives.

Notably, my understanding of critical spiritual literacy requires a deeper exploration of the assumptions embedded in dominant and Indigenous cosmologies and the consequences of taking for granted Euro-Christian discourses and ways of reading the sacred as normative and universal.

Distinctive Features of the Book

The Souls of Yoruba Folk: Indigeneity, Race, and Critical Spiritual Literacy in the African Diaspora offers a number of distinct features that make this book unique. The first is the central organizing concept of critical spiritual literacy, used as an analytical tool that is woven throughout the text. A primary objective in this book is to shift our notions of what spiritual literacy is, and what it does by prioritizing a critical component: equity and oppression. In so doing, spiritual literacy or reading the sacred is approached from a critical position where social justice is the ultimate goal.

The second distinct feature in this text is the theme of Yoruba Indigeneity as a social positioning that is counterhegemonic. This is where more nuanced and spiritually anchored understandings of the complex religious lives of Yoruba migrants are engaged. This theme is taken up through the humanizing frameworks of Indigenous cosmology, Black/African feminisms, and anti-colonial theory, where the intersections of race, class, gender, age/seniority, and African diasporic Indigeneity are heavily relied upon. With the specific realities of Indigenous Yoruba in the African diaspora as the example, "diasporic Indigeneity" is a new and distinct term articulated by the author to counter simplistic understandings of Indigeneity as a social location, thereby lending itself as another unique feature of the book.

Third, as mentioned above, Ifa/Yoruba Indigenous spirituality is arguably the most popular expression of Indigenous African spirituality in the African diaspora (M. Asante, personal communication, 2012). Nevertheless, there are few studies that focus on Yoruba Indigenous spirituality, as expressed and understood in the lives of the more recent Yoruba diasporic communities. While much scholarship has been written on African descended communities that openly practice the Yoruba derived religion of Ifa in the diaspora and on the continent, the focus has been on this community of believers in terms of their shared spiritual practice as followers of this faith (Abimbola, 1976, 1977, 1997a; Bascom, 1969; Bascom & Herskovits, 1959; Emanuel, 2000; Soyinka, 1959, 1981, 2012). However, in this book, the reverse is done: The entry point and focus is the more recent Yoruba ethnic and linguistic group in the African diaspora, calling attention to how they understand and engage with their Indigenous Yoruba spirituality (Ifa).

Fourth, whereas literature on the African Atlantic diaspora is fairly developed and well established in the United States, Britain, and the Caribbean, the African diaspora in Canada is often overlooked and rendered invisible. This book contributes to the sparse yet growing scholarship on the specificities of the Canadian experience in the African diaspora. It uniquely closes the gap between African (Yoruba) continental Indigeneity and how these Indigenous spiritual knowledges are taken up in the African diaspora, namely the under-researched area of contemporary African diasporic life in Canada. The absence of such studies has also meant a silence around how Yoruba Indigenous culture is affected and reconfigured in the face of balancing transnational familial and economic demands from within the onerous context of colonial oppression. Little scholarship investigates how African migrants negotiate their Indigenous identities in colonized spaces that are not Indigenous to them, making this project unique for its relevance to both the diasporic Yoruba featured in this book, and those who are socially positioned in similar ways. Consequently, the book's focus on this significantly under-analyzed area will bridge this gap and provide a meaningful contribution to advancing knowledge on spirituality and Indigenous thought among people of African heritage, both from Africa and its diaspora.

Additionally, the author's discussion of spirituality in the contexts of schooling, education, and teaching offers insight into what equitable pedagogy and research might entail and how it can be applied, practiced, and taught in the classroom.

Finally, in challenging the subjugation of African Indigenous ways of knowing, a space for critical dialogue regarding the empowering spirit of these knowledges is opened up. In other words, this book serves as an example that illustrates the vital roles that African spirituality and Indigenous thought play in our decolonization, self-determination, and liberation.

A Call to "Learn to (Re)member the Things We've Learned to Forget"

The Souls of Yoruba Folk is about garnering a more nuanced understanding of the complex lives of Yoruba migrants in the African diaspora, a diaspora that exists within the geopolitical space of dominant Euro-Canadian culture. I have argued that African spirituality and Indigenous knowledges play a vital role and function in the lives of Africans, and are therefore worthy of our study and attention. I have also argued that this project is both a personal and political effort to de-pathologize African spirituality and shift the popular imagination (African and otherwise) in its profound reliance on re-circulated colonizing scripts about African Indigenous spirituality, to ones that are affirming, nuanced, and grounded in an African-centered, feminist, and anticolonial politic. Of equal significance, this book aims to retrieve and center the voices, experiences, and knowledges of Africans: as knowledge producers; as our own explorers seeking and (re)discovering the meaning, value, and power of our Indigenous heritage on our own terms; as grounded in our own cosmologies and ways of knowing, diversely unified as they are. This book is the author's small contribution in the form of a call ... to the Souls of Yoruba folk, to "learn to (re)member the things we've learned to forget" (Dillard, 2012). Grounded in the ethos of substantive equity, it is a call to acknowledge and celebrate African humanity as worthy, valid, and equitable.

On Double-Consciousness and Warring Ideals

> [T]he Negro is a sort of seventh son [daughter], born with a veil, and gifted with second sight in this [North] American world—a world which yields him no true self-consciousness, but only lets him see himself through the revelation of the other world. It is a peculiar sensation, this double-consciousness, this sense of always looking at one's self through the eyes of others, of measuring one's soul by the tape of a world that looks on in amused contempt and pity. One ever feels his two-ness,—a [North] American, a Negro; two souls, two thoughts, two unreconciled strivings; two warring ideals in one dark body, whose dogged strength alone keeps it from being torn asunder ... In those sombre forests of his striving his own soul rose before him, and he saw in himself,—darkly as through a veil; and yet he saw in himself some faint revelation of his power, of his mission. He began to have a dim feeling that, to attain his place in the word, he must be himself. (W. E. B. Du Bois, 1961: 16–17, 20)

In Yorubaland, Nigeria, Sade was married, with two children and a comfortable job as a civil servant with the Nigerian federal government, a position she held for 11 years. However, due to increasing political problems and marital troubles, at 29, Sade decided to leave Nigeria for Canada and applied for residency as a political refugee. Her attraction to move to Canada came from hearing of its multiculturalism, where one could freely live and practice one's culture; not having heard too many bad things about Canada; and finally, knowing fellow Nigerians who had also moved to Canada and reported it as a family-friendly country. Upon arrival to her new home, Sade found that her accent gave her problems integrating and finding employment that she knew she was qualified for. She spent almost 10 years working in factories and jobs of the like before being hired for a data entry position due to her typing speed. To make herself more employable, Sade obtained bachelor degrees in social work and sociology to add to the diploma in secretarial studies she had acquired in Nigeria. In addition to linguistic racism and discrimination, Sade found the lack of respect for elders—where younger people looked their elders directly in the eyes, and called them by their first name—alongside the different (i.e., limited) notions of family and community as the most challenging adjustments in making Canada her new home, adjustments, Sade admitted, that she was still struggling with given that "Back home, it takes a whole community to build up the kids. It's not like it's just me and my wife or husband, it's an extended family; so people are always there for each other so you [as a parent] are not worried too much."[1] Drawing from her upbringing in Nigeria, Sade also developed a reputation entertaining local communities in Toronto as a storyteller, actor, and dancer, particularly in the Yoruba, Nigerian, and African Canadian communities.

It would be another eight years, in 1999, before Sade's sons joined her in Toronto, Canada. Dele, the younger of the two boys—who was also interviewed for this book—came to join his mother when he was 13 years old. Similar to his mother, Dele found that his "thick African accent [and] being made fun of by teachers"[2] was one of the most challenging of the difficulties faced in adapting to his new life in Canada. Dele also felt that it was "more community oriented back home in Nigeria."[3] According to Dele, "there was like—I don't know about now, but the electricity problems[4] has its good benefits in terms of gathering together when the light is off and storytelling with children and people around the compound … I miss that."[5] For Dele, his identity as Black was important and meant that one could show their pride by "avoiding certain statistics like being seen as violent and trying to stay out of trouble … and not looking at yourself as inferior. It's negative but that's just the way it is."[6]

For more than 20 years Sade Oriola has lived in Toronto, Ontario, Canada, working and raising her two sons, both of whom have lived in Toronto for almost 15 years now. Sade and her family have been trying to combine and reconcile the knowledge, culture, and upbringing of their Indigenous Yoruba homeland with Canadian culture, while making sense of the experiences garnered from this. Such attempts to "combine," "merge," and "reconcile" two different cultures (one dominant and oppressively White supremacist, and the other inferiorized by such dominance and repression) is what W. E. B. Du Bois has referred to as the "double-consciousness" of "spiritual strivings" for Black folk. For Sade and her son Dele, I am interested in the unreconciled spiritual strivings of their Indigenous Yoruba identity—namely, their spirituality. I am interested in the warring ideals of the two souls Black folk possess in Eurocentric Western culture. For Indigenous Yoruba living in Canada such as Sade and Dele, one soul is layered as African/Black/Yoruba/Indigenous and the other, also layered, is Canadian/Western/colonized. In essence, I am interested in how these warring ideals are entangled and experienced in the "one dark body" of the Yoruba diaspora. In essence, I am interested in the souls of Yoruba folk.

These interests were highly influenced by the life and groundbreaking work of renowned intellectual W. E. B. Du Bois, whose scholarship and activism steered me toward embracing education not only as a profession but also in understanding education as a life of service. It was his especially seminal work, *The Souls of Black Folk*, which sparked something inside of me, then a graduate student who, while passionate about and versed in equity concerning issues of race, class, gender, etc., felt there was something else that needed exploring but could not articulate what *that* was. Until one day, while perusing various titles in one of my favorite Black bookstores, I came across *The Souls of Black Folk*. After purchasing the book, I read and reread it within one week. It was Du Bois' focus on the combination of race and religion; his passionate argument of Black folks' two warring souls; and his clear love for Black folk (well-defined in the eloquent brilliance of his writing and argument) that made these souls, *our souls*, reconciled. This is what in large part gave me permission to say *yes*. The religion, the spirituality, the souls of Black folk *are important*, *are valuable*, and are a powerful story that needs to be told. They *are* worthy of academic study. It was this, combined with a graduate course I took on Decolonization and Indigenous Knowledges, which led me, a Yoruba woman in Canada, on the journey to exploring and better understanding the souls of Yoruba people.

Beginning with the premise that spirituality and indigenous knowledges are of vital significance and critical function in the lives of people of African[7] heritage, the central focus of this text entails an exploration of the spiritual lives and experiences of Africans of Yoruba descent in Canada, such as Sade and her son Dele. For Africa and Africans, the question of spirituality is deeply bound up within a history of centuries of colonialism, and is therefore entwined in questions of power, knowledge, historical displacement, and cultural genocide. However, upon closer examination of this history, spirituality is also bound up with/in a complicated matrix of social agency, an individual and collective negotiation with and resistance to colonial domination. For Africans living in Canada, such complexities are heightened by dominant Eurocentric constructions of African identities, particularly as an inferior monolith. Of particular relevance is the racist pathologization of African Indigenous knowledges (Asante, 2009, 2003; Dei, 2000; James, 1993a; Morrison, 1984). Contrary to colonialist and marginalizing constructions of these ways of knowing, Indigenous knowledge systems are—and have been—crucial sites of empowerment and resistance particularly for those who reside in dominant Eurocentric contexts (Dei, 2000).

The Souls of Yoruba Folk: Indigeneity, Race, and Critical Spiritual Literacy in the African Diaspora examines this inequitable contention between colonizing/Eurocentric knowledge and Yoruba Indigenous knowledges by investigating how first- and second-generation diasporic Yoruba utilize, negotiate, and make meaning of their Indigenous spirituality within the geopolitical space of dominant Euro-Canadian culture. Drawing from African-centered theoretical frameworks such as Yoruba Indigenous literature, African/Black feminisms, and anticolonial theory, the experiences of diasporic Yoruba are situated within a sociohistorical and cosmological context to effectively examine the impact of dominant culture on Yoruba migrants' constructions of their Indigenous cultural heritage. However, this book also highlights how Yoruba peoples in the African diaspora strategically re-member, that is, how we use our Indigenous spiritual knowledges as decolonizing tools of navigation, subversion, and resistance to oppression in the purportedly multicultural space of Canada. In essence, this text is also about "learning to (re)member the things we've learned to forget" (Dillard, 2012).

Here, I explore the extent to which Eurocentric colonialism has amnesically marginalized and constrained Yoruba folks' abilities to re-member, secure, and sustain self-affirming constructions of Yoruba Indigenous identities, as well as the implications and decolonizing possibilities of such realities. Hence, this exploration is informed by the following questions. How do Yoruba Indigenous knowledges inform the diasporic lived experiences of Africans of Yoruba descent in Canada, and vice versa? What are the challenges of learning and honoring these Indigenous knowledges (particularly the spiritual paradigm that I argue frames these knowledges) within in a context that pathologizes Yoruba (African) Indigenous spirituality? What are the subversive and emancipatory possibilities of using Yoruba Indigenous knowledges? For example, how can we begin to develop an empowering lived pedagogy of Indigenous African identities from within the lived context of dominant Euro-Canadian culture?

These questions were gleaned from my experiences concerning the complex challenges I encountered in embracing my Indigenous spirituality and culture as a second-generation Yoruba-Canadian in the African diaspora. Accordingly, I hope the reader will also witness my attempt to (1) develop and contribute to critical social theory about Yoruba (African) Indigenous knowledges; (2) engage in in-depth learning and discussion of Yoruba lived experiences and understandings of Yoruba cosmology from within the larger contexts of Euro-dominant culture and; (3) open up a space that engages critical dialogue about Yoruba Indigenous identities and knowledges that are more affirming and accessible. Particular attention is required in public institutional spaces where these knowledges have been rendered invisible, silent, and bound up in racist colonial constructions.

The Personal Is Political and the Political Is Personal

This book was informed by my life and educational experiences and how the legacy of colonialism has shaped my relationship to Yoruba cosmology, for example, how my access to Yoruba cosmology has been fragmented and blocked. Drawing from the feminist ethos that our personal lives are entrenched with the inescapable politics of our social world, I felt the research also required the exercise of writing myself into this work. I do so for reasons that are powerfully articulated by African American feminist scholar Cynthia Dillard (2006):

> Given our training and academic preparation as researchers, teachers, and scholars, we all too often gaze and burrow into the lives of others in pursuit of our projects, but too seldom turn the gaze on ourselves, our work and the reasons we do what we do. This book is an attempt to bring a critical gaze to my own academic life and work, embracing the messiness, tensions, and complexities involved in writing myself and other African women scholars into the text … For me, writing that is both theoretical and biographical can also provide a glimpse of ourselves and the transformational power of the spirit in our lives as African American [and Canadian] women academics, illuminating the spirituality that is all too often rendered invisible or insignificant in (white) academe by virtue of our race, gender, and other identity positions and the "isms" others embrace. (xi)

Hence, I came to this work by virtue of my life experiences of oppression, violence, and amnesia that were persistently interwoven with a strong sense of displacement and loss around not knowing my Yoruba culture and identity, as well as feeling cut off from important parts of me that seemed so inaccessible. However, I was also deeply anchored by an equally strong sense of my Blackness, which was shaped largely by diasporic Africans who had a similar yet different history of struggle, trauma, and resistance in the "New World." As I continued to make connections with Africans who also had a passion for African spirituality, and yearned for but did not have the language to access it, my specific location and history as a Yoruba (African) woman who could point to and claim a particular people, language, and land on the continent of Africa became evident. I became curious about developing a deeper historical understanding of the varied ways Africans have resisted, challenged, and negotiated our shared histories of colonialism. Exchanging these differences alongside our similarities helped me to realize that I wanted to know more.

This led me to an opportunity and space in which to grapple with, explore, and re-member my Yoruba identity. Through many conversations with elders in the Yoruba community, and through attending many Yoruba functions in Toronto, I realized that the particular forms of Yorubaness expressed were more often than not anchored in Christianity and, to a lesser degree, in Islam. Similar to my upbringing in Canada, while many first-generation Yoruba Canadian migrants would speak Yoruba to each other (their peers), I found it interesting that they spoke English to their children and that even if they did speak Yoruba to their children, their children almost always replied in English. I realized there was a pattern to this and wanted to understand why this was the case. Why was it that normal Yoruba identities were overwhelmingly seen as the colonially derived ones? The most obvious examples to me here were religion and/or spirituality and language. I wondered what had happened to Indigenous Yoruba spirituality. Why was it that in conversation with many community elders, our Indigenous spirituality was repeatedly seen as nonnormative, and why was it largely spoken about with deep disdain, shame, or almost always dismissed as something to not talk about? Additionally, why were so many parents not speaking the Yoruba language to their children, or, conversely, why were so many children of Yoruba migrants not able to speak, much less read or write, Yoruba? Repeatedly, at community outings, family and friends' functions, and church, I noticed this same pattern emerged. While these were my personal anecdotal observations, I wanted to know: Were people in the Yoruba community really for the most part not—or at least not openly—embracing and practicing Yoruba spirituality? And why did so many first-generation Canadian children, such as myself, not speak much, if any, Yoruba at all? Although highly valuable from an academic standpoint, my experiences were largely anecdotal assumptions that had not been substantiated with scholarly research.

However, as a young Black woman who was raised in Canada since the age of three, I also knew about racism, sexism, and classism and had experienced the traumatic realities of these forms of oppression on a daily basis. Coming from a working-class family, I knew what it meant to not be able to afford to buy my lunch or have the same latest brand name clothes and toys as my White peers. Both myself and my siblings began part-time jobs at early ages in our adolescence, and in doing so, we were partially responsible for family expenses. I knew what it meant to either be ignored in a store because it was assumed that I could not afford the merchandise, or to be followed closely because it was assumed that I would steal the merchandise. I also knew what it meant to be in Canadian public schools since the age of three and never experience being taught by a Black teacher until graduate school, because the principal and school officials at all three levels (elementary, middle school, and high school) claimed that "there weren't any qualified Black teachers to hire." I knew what it meant to come home from work on a school night and have a dark car following me, only for the window to be rolled down, often exposing an older white man asking me if I wanted a ride. I knew what it meant to see young Black males in my high school be used for basketball competitions while getting credited for classes they did not have to attend because they were going to be basketball stars. Yet as it turned out, they were merely being used to bring in a quiet profit for the school coach from corporate sponsors. I also knew I never learned about African/Black people in school and if I did, they were racist stereotypes such as violent savages, happy slaves, or anonymous bodies stricken with disease due to famine or civil war, always inevitably in need of Western aid. Yes, despite the national rhetoric of Canada as a tolerant and peace-loving country, I knew that my experiences as a young Black woman were contrary to this. Racism and other forms of oppression were alive and well in this multicultural country.

Hence, this book became a space where I could investigate the problem of what, to me, seemed to be deep-seated patterns of Yoruba cultural discontinuities within a larger context of deep-seated anti-Black racism and the many other forms of institutional oppression that exist in Canada. Being in graduate school allowed me a space to remain cognizant of and merge these experiences with the historical realities of how "Canada"[8] came to be. In being provided a space to continue asking deeper and more critical questions, I was able to marry them with my experiences in terms of exploring what it means to be woman, colonized, Black, decolonizing, African, Canadian, and to explore how these layers figure into the shared Yoruba identity I have with the 16 participants. However, in highlighting this shared, but not singular, Yoruba identity, it was clear that I also needed to be cognizant of difference and how aspects of our identities contribute to differing standpoints and social locations (Amadiume, 1997, 2000; Carty, 1996; Collins, 1990; Lorde, 1984). In other words, both the shared and different experiences that are structured around race, age, gender, class, and generation serve the function of nuancing the exploration of Yoruba Indigenous identities in African diasporic contexts.

Ultimately, my particular social location is paradoxically both as an Indigenous "insider" and a Western researcher. Feminist sociologist Dorothy Smith (1987) asserts the importance of challenging notions of objectivity in research:

> The practice of objectivity in the social sciences allows that science to detach its corpus of statements from the subjectivities of those who have made them. It has very little to do with the pursuit of knowledge. (33)

I would like to extend Smith's discussion of the impossibility of objective research to the impossibility of objective identities, as we all have a particular perspective, social location, and standpoint from which we know and understand our world, lives, and experiences (Collins, 1998; Dillard, 2012; Lorde, 1984). However, my position as the researcher in this project posed additional challenges and tensions around power, namely academic training and exclusiveness, and the politics of knowledge production. This dilemma has often been understood as that of the insider/outsider (Carty, 1996; Collins, 1990; Smith, 1999). In *Decolonizing Methodologies: Research and Indigenous Peoples*, Linda Tuhiwai Smith (1999) writes:

> There are a number of ethical, cultural, political and personal issues that can present special difficulties for indigenous researchers who, in their own communities, work partially as insiders, and are often employed for this purpose, and partially as outsiders, because of their Western education or because they may work across clan, tribe, linguistic, age and gender boundaries. Simultaneously they work within their research projects or institutions as insiders within a particular paradigm or research model, and as outsiders because they are often marginalized and perceived to be representative of either a minority or a rival interest group ... More often, however, I think that indigenous research is not quite as simple as it looks, nor quite as complex as it feels! If I have one consistent message for the students I teach and the researchers I train it is that indigenous research is a humble and humbling activity ... Indigenous researchers are expected, by their communities and by the institutions which employ them, to have some form of historical and critical analysis of the role of research in the indigenous world. In general, *this analysis has been acquired organically and outside of the academy*. (5; emphasis added)

What Smith highlights are the politics of the extent to which Indigenous knowledges are marginalized and devalued within academic spaces. Ultimately, this is because the knowledge that researchers who identify as Indigenous draw upon have been consigned as inferior and overwhelmingly exist outside the oppressive and exclusive terrains of the academy. I therefore write and produce this research with a clear social and political project of decolonization to affirm Yoruba Indigenous identities and knowledges, particularly in colonizing and imperial contexts. To this end, it is these multiple and overlapping positionalities, that is, insider/outsider, Indigenous yet Canadian, Yoruba, woman, Black and decolonizing, that I bring to this book.

Organization of the Book

Using the African Indigenous ritual of call and response, each chapter of this book focuses on distinct but connected themes as critical moments in the author's call to Yoruba folk. In Chapter 2, a call is made to reimagine how notions of Indigeneity are taken up so that the unique realities of diasporic Africans can be accorded a space to theorize the particularities of their experiences. Diasporic Indigeneity is offered to explicate the social positioning of Indigenous people living in and navigating the complexities of a land that they are not indigenous to. Inevitably, this chapter sets the theoretical backdrop for the reader, so as to ground them in an understanding of the important ideas, theories, and concepts that will aid in their journey of developing more nuanced readings concerning the spiritual lives of Yoruba diasporic communities.

Chapter 3 spotlights each of the Yoruba elders and youth featured in this book, and through discussion of each individuals' spiritual consciousness and connection to Yoruba Indigenous culture, I introduce the reader to the souls of Yoruba folk. Included in this chapter is an examination of Yoruba community members' understandings of spirituality, religion, and how this faith figures for them.

At the heart of Chapter 4 is a discussion of the sacred through the Yoruba Orisa, Esu. In particular, this chapter offers critical inquiry into how Yoruba elders and youth in the African diaspora understand this deity to figure in their spiritual lives. Chapters 3 and 4 are closely linked by an overarching theme of spirituality and religion as sites of contestation. The metaphor of the closet, covert silences and the denial of Yoruba Indigenous spirituality as sacred are explored through critical examination of the Orisa Esu as a symbolic representation of Yoruba spirituality and how this deity is constructed by diasporic Yoruba in Canada. In these chapters, the nuanced contradictions and challenges of embracing one's Indigenous spiritual identity in dominant Eurocentric Canadian culture are discussed. Woven throughout is the author's gentle call encouraging diasporic Yoruba to come out of their spiritual closets and affirm their Yoruba Indigeneity.

Chapter 5 examines the perils and paradoxes concerning the social category of age, or what can be loosely translated as seniority in Yoruba Indigenous terms. Yoruba elders and community members' conceptions of age, the role of elders, and respect for elders are analyzed and explored as they culminate in seniority as a salient spiritual and social category in Yoruba life. Drawing from both elders' and community members' narratives and the wisdom of Yoruba worldsense (Oyewumi, 1997), the author suggests that seniority is both a contested and contextual social category that can change, shift, and be contradictory. In this chapter, the author is making a call to the souls of

Yoruba youth in particular, to explore and (re)member what they've learned to forget (Dillard, 2012) by facing and countering any fears they may have of their Indigenous spirituality.

Chapter 6 brings the reader full circle with the author asking for her call to be answered: answered by herself, answered by the Yoruba community, answered by the larger African community, and, ultimately, answered by humanity. In this chapter the implications of *The Souls of Yoruba Folk* are discussed: for example, mapping how the framework of critical spiritual literacy has better helped us understand the Yoruba worldsense and the contradictions embedded in the spiritual lives of Yoruba migrants; mapping a nuanced understanding of what it means to be diasporic and Indigenous at the same time, on land that you are not Indigenous to; mapping what it means to be in a context that requires a devil, to then characterize a salient figure in Yoruba worldsense as that demonized figure; and mapping the consequences of this transposition. Lastly, remembering that embedded in this call is a nuanced articulation of what these complicated realities and experiences do to a people's soul, as individuals and as a collective.

Finally, Chapter 7 closes the circle with an open letter to a group of people who haven't been specifically referenced in this book, yet are deeply implicated: teachers. Featured in this closing chapter is a call to teachers to recognize and understand the critical role they play in shaping and influencing the souls of Yoruba folk, namely those of the youth. Given the high regard of children as indispensably valuable in the African community, the power teachers have, in terms of the influential role they play in Black/African children's lives, is not to be understated. This chapter is about a politics of hope and the responsibility educators have in engaging spirit in their pedagogical practice. Teachers are a powerful example of a group of people who may or may not be Yoruba, may or may not be African, yet are undoubtedly involved in knowledge production and education, how it is constructed and how it is taught. In this way, teachers are called to an important charge: to rise to the challenge of teaching from a critical place where social justice and equity are centered and central to one's practice.

In short, the call is one of forward movement. Armed with the wisdom and beauty of one's Indigenous knowledges, it is a call that beckons an exploration of empowering possibilities and meeting whatever challenges may be borne of this, all for the purpose of envisioning and creating a more equitable future for Africans, and for humanity as a whole.

Notes

1. Sade Oriola, interview, 10 March 2007. See the Appendix and the section titled "In Dialogue with the Souls of Yoruba Folk: Oral Interviews" for more details on the interviews.
2. Dele Oriola, interview, 10 March 2007. See the Appendix and the section titled "In Dialogue with the Souls of Yoruba Folk: Oral Interviews" for more details on the interviews.
3. Ibid.
4. For the past several years (four decades and counting) Nigeria has been struggling with providing consistent power and electricity to its citizens. So much so that the national organization responsible for this service in Nigeria, the National Electricity Power Authority (NEPA), has been humorously renamed as "**N**ever **E**xpect **P**ower **A**lways" among its citizenry. However, NEPA is now formally known at the Power Holding Company of Nigeria (PHCN).
5. Dele Oriola, interview, 10 March 2007. See the Appendix and the section titled "In Dialogue with the Souls of Yoruba Folk: Oral Interviews" for more details on the interviews.
6. Ibid.
7. I include in the term "Africans" both persons whose countries of birth were African and those whose racial, cultural, and ethnic origin is from the African continent. I also use the terms "African" and "Black" interchangeably, approaching them as politicized identities to highlight the specificity of the commonly subjugated and racialized social, political, and economic status of Africans due to colonialism, imperialism, and transatlantic slavery.
8. I use quotations to highlight the paradox that "Canada" is a "real" material and geographic space, yet at the same time a space and land that remains marked, reflective of colonial hierarchies and power imbalances. Canada is an occupied land belonging to the first/Indigenous peoples. Accordingly, I also use quotations to highlight the layers of this land, symbolically denoting this geographical space's Indigenous name: Turtle Island.

· 2 ·

THEORIES OF DIASPORIC INDIGENEITY AND BLACK FEMINISMS

Living and Imagining Indigeneity Differently

The Souls of Yoruba Folk is anchored in three theoretical frameworks: Indigenous knowledges, anticolonial theory, and Black/African feminisms. I draw from and build on these theoretical models of analysis to critically contextualize the experiences of Yoruba peoples in diasporic and Euro-dominant contexts. Indigenous literatures and anticolonial and African/Black feminist frameworks allow for a more nuanced and critical reading of how issues of race, class, spirituality, gender, language, religion, and especially notions of Indigeneity *interlock* in the lives and experiences of Yoruba peoples in the diaspora.

Within the anticolonial discursive framework, there is a particular focus on the term "Indigenous" as a vitally significant concept in anticolonial thought. I argue that this term needs to be revisited, extended beyond existing ideas, and critically interrogated where diasporic Africans are concerned, so that critical spiritual literacy as a conceptual tool can be effectively understood and applied in this book. I maintain that Indigeneity (or Indigenous identities) need to be imagined differently so that the unique positionings of especially diasporic Africans can be accorded a space to theorize the particularities of their experiences. In other words, there is a need for a shift in how notions of Indigeneity are taken up so that they are not imagined as singular, in the way that those who often work from exclusively Eurocentric perspectives do. Hence, more flexible approaches with Indigeneity need to be engaged because this concept is often taken up to exclude diasporic African identities. Instead, this intellectual shift in notions of Indigeneity needs to include a variety of Indigenous peoples' experiences so that Indigeneity or Indigenous is engaged in ways that allow a discussion of something I term "diasporic Indigeneity." Moving on to African/Black feminist theory, the key argument is that gender in Indigenous and especially the Yoruba context differs from Eurocentric understandings. There is, therefore, the need to eschew these dominant notions of gender that mask themselves as universal, specifically in one's thinking and approach to Indigenous peoples and identities.

"Indigenous" is a relatively recent term that emerged in the 1970s out of the American Indian Movement (AIM) and the Canadian Indian Brotherhood (CIB) (Smith, 1999). It came into being to give a common name to both those who exist outside the colonial domain as well as those who were colonized (Narogin, 1995). Trevor Purcell (1998) provides further background regarding usage of the term as one that is largely self-applied because it carries less condescension than words such as "primitive" or "tribal" (259). However, this concept is not unproblematic and has been seriously challenged primarily by those who work from postmodern theoretical frameworks because it may appear to homogenize unique and distinct peoples that have different experiences under imperialism (Smith, 1999). Such opposition has not come from Indigenous peoples themselves (Battiste & Henderson, 2000; Dei, 2000) but rather has originated with scholars whose work relies on Eurocentric frames of mind and analysis as the sole loci of legitimate knowledges. For Indigenous peoples, usage of the term "Indigenous" is a *collectivizing* political and social strategy that emerged out of anticolonial social movements such as AIM, CIB, and Red Power, all of which were strongly influenced by the American Black Power and Civil Rights Movements of the 1960s. Furthermore, "Indigenous" is employed as an umbrella term, or way of including common experiences of colonialism across various communities, language groups, and nations. Using the term need not deny that each group's distinctiveness and unique experiences with imperialism will also be given serious attention and acknowledgment (Smith, 1999). Ultimately, "Indigenous" is a multilayered term informed by a multiplicity of experiences where both sameness and difference exist simultaneously; it is not a singular category, experience, or identity. Hence, my particular understanding of Indigeneity may not relate to or include the experiences of all of those who speak from an Indigenous positionality and identity. For some, they are better suited, or more comfortable, focusing on the variety of issues that surround Indigeneity rather than placing emphasis on what often can be limiting definitions. I elaborate on this further in the following section. It is important to note that while I do recognize that a multitude of claims and oppositions to Indigenous as an identity concept exist, it is not my intention nor goal to resolve the conflicts and debates around this term. Instead, I offer my multilayered working definition of how this term is used and conceptualized by me, and how it specifically applies to Yoruba (African) peoples.

First Layer: On Lands of Origin

The first layer of my thinking comes from Mudrooroo Narogin's (1995) conception of Indigenous to "simply mean originating in or from a country [or land]" (7). For Yoruba peoples, this land is in the south western region of what is now known as Nigeria, as well as the neighboring borders of Togo and Benin. While according to archeologically documented knowledge, Yoruba presence in this region dates as far back as A.D. 800, Indigenous Yoruba oral stories designate this region as the site of the birth of humankind. However, the culture, language, spirituality, and worldview of the Yoruba are not bound to this one geographic space, but have traveled with Yoruba peoples in the myriad directions they have moved and migrated outside this region. For example, forced removals such as the European slave trade, which resulted in a large number of Yoruba peoples being dispersed to areas that include (but are not limited to) other parts of Africa, as well as other geographic regions such as the Americas, the Caribbean, and the continent of Europe, have inflected Yoruba identities with other lived experiences in other geographic spaces.

Second Layer: The Significance of a "Yoruba Worldsense"

The second layer of my use of the term "Indigenous" concerns conceiving of cosmology or worldview as foundational, and borrows from George Dei's (2000) conception of Indigenous knowledges as dynamic, experientially based, holistic, and relational in the sense that the interwoven nature of the physical and metaphysical realms of Yoruba life are cosmologically anchored and common knowledge. In Dei's (2000) words, "Indigenous epistemologies are grounded in an awareness and deep appreciation of the cosmos and how the self/selves, spiritual, known and unknown worlds are interconnected" (115). This is in keeping with the Indigenous Yoruba philosophy of maintaining links to ensure that connection between these worlds remain active and unbroken. For Yoruba peoples, communication occurs through myriad rituals and practices that use and evoke all bodily senses. Yoruba feminist Oyeronke Oyewumi (1997) explains the inadequacy of the term "worldview" as a synonym for cosmology in Yoruba contexts:

> The term "worldview," which is used in the West to sum up the cultural logic of a society, captures the West's privileging of the visual. It is Eurocentric to use it to describe cultures that may privilege other senses. The term, "worldsense" is a more inclusive way of describing the conception of the world by different cultural groups ... [and] will be used when describing the Yoruba or other cultures that may privilege senses other than the visual or even a combination of senses. (3)

Overemphasis on the visual is also problematic for Yoruba Indigenous discourses because the Yoruba conceive of spirituality and/or spiritual forces as largely inaccessible to the human eye. Hence, the term does not fully reflect the complexity of Indigenous Yoruba culture and how life is understood from this multisensed position. Cosmology or worldsense is a foundational locus of Indigenous peoples and their knowledge systems. This is so because a people's worldsense maps out how they experience and understand their world. In the Yoruba context, cosmology is complementary (Olajubu, 2003; Oyewumi, 1997; Soyinka, 1976), and interconnectedness is central, where *Orun* (the otherworld), *Aiye* (the physical world of living human and other beings), and *Ile* (the earthworld) are all interdependent and cannot exist on their own. *Orun* is inhabited by the Supreme Being, *Olodumare*, who is also known as *Eleda* (the Creator) and *Olorun* (literally meaning owner of the skyworld); the *Orisa*, many of which once walked the earth as human beings with mystical powers, were then deified through death. The lives of the Orisa are continued through the supernatural powers and prowess of various forces of nature such as water, wind, land/earth, fire, thunder/lightning, and the forest/trees. The Otherworld is also inhabited by a number of spirits such as *Egungun* (Ancestors), *egbe* (mirror or spiritual half/halves on the otherside), *Ebora/Iwin* (a small supernatural being with magical powers[1]), and *Ara Orun* (beings of the otherworld, including the unborn). The forces of *Orun* are always influencing and in communion with those who inhabit *Aiye*. The people of the physical world (aiye) will one day die and simultaneously become one with *ile* (the female earth/soil) because it is our last resting place, as one moves on to the spirit world. The earth and its connection to the dead are so sacred to Yoruba peoples that relatives of recently buried loved ones take small portions of soil from their grave and use it to swear oaths (Abimbola, 1997a: 68). These ties illustrate the interconnection between the living and the dead through the power of nature and earth. As is common in many Indigenous cultures, the Yoruba also hold nature and one's environment to be sacred. This is reflected in our belief in deified ancestors, the Orisa, and spirits that are associated with natural phenomena such as mountains, hills, earth, rivers, lakes, the ocean, trees, and wind (Awolalu, 1979: 45). An example of *ile* as sacred in Yoruba life and cosmology is explained by historian J. Omosade Awolalu (1979):

> The earth is venerated in Yorubaland because it is believed to be inhabited by a spirit. The Yoruba attach great importance to the earth. In creation, the myth says, earth was spread on the face of the deep, and land appeared. Furthermore, Obatala used clay to mould man before Olodumare gave him breath. When a new born comes into the world, the first landing place is the earth; when a man grows old and dies, he is buried in the earth. The earth supplies food for human consumption, and so it keeps life going. From the Yoruba point of view, an element which has such manifold and useful functions must have a spirit dwelling there ... Since most of the Yoruba depend on agriculture for their sustenance, and crops are grown in the soil, Ile (the earth) receives special sacrifice at the time of planting and harvesting, almost in the same way as Orisa-oko does. And since the corpses of the ancestors are buried in the earth and there are powerful spirits dwelling therein, the Yoruba have the habit of pouring the first drop of any drinks on the ground and of throwing some portion of food to the earth before they drink or eat in order that the spirits may drink and eat first. (45)

Third Layer: On Different Belongings and Relationship to Land

The third layer of how I conceptualize Indigeneity concerns the matter of land and how it is understood as the definitive marker of who can officially be identified or counted as Indigenous. Dr. Erica-Irene Daes, the chair of the United Nations Working Group on Indigenous Populations, places particular emphasis on land or ecology as "the central and indispensable classroom" for the teaching and generational transmission of Indigenous knowledge systems (Daes as cited in Battiste & Henderson, 2000: 41). Many Indigenous scholars and intellectuals such as Ward Churchill, Linda Smith, Andrea Smith,

Ngugi Wa Thiong'o, Ama Ata Aidoo, Ifi Amadiume, Toni Morrison, Wole Soyinka, Wande Abimbola, and others have stressed the significance of land for the self-determined development and sustenance of Indigenous peoples and their systems of knowledge. On this matter, Aboriginal scholar Marie Battiste (2002) writes:

> Indigenous knowledge is also inherently tied to land, not to land in general but to particular landscapes, landforms, and biomes where ceremonies are properly held, stories properly recited, medicines properly gathered, and transfers of knowledge properly authenticated. Ensuring the complete and accurate transmission of knowledge and authority from generation to generation depends not only on maintaining ceremonies, which Canadian law treats as art rather than science, but also on maintaining the integrity of the land itself. (13)

I do not dispute that land, or rather particular lands, are central to many if not all Indigenous peoples and the knowledge systems they maintain. Rather, it is the manner in which land is spoken of that is problematic for me here, especially where many diasporic African peoples are concerned. It is often implied or assumed that Indigenous peoples are current residents of their countries of origin. However, just as often, they may have been displaced and pushed off their distinct territory through colonization. While imperial or colonialist settlers often forcibly remove Indigenous peoples from their territories, often they remain residents of that same larger physical/geographic space or land mass—as is the case with Aboriginal peoples in the Americas, Australia, the Pacific region, as well as many South African Indigenous peoples. Trevor Purcell (1998) reminds us that to be considered Indigenous one must be residing on and/or in relatively close proximity to one's ancestral territory. However, what has not been given due attention—particularly in academic as well as Indigenous grassroots activist circles in the Americas—are the Indigenous populations that have been physically displaced *off that particular land mass*. This is the case with the hundreds of millions of Africans who were enslaved by Europeans, or the current movements of continental Africans who follow global capital as a result of globalization (which in itself is simply a new form of imperialism). In this sense, it is one's *relationship* to land that is critical, and this does not necessarily manifest as a physical marker. Yes, one's Indigeneity is undoubtedly tied to a distinct ancestral land or territory, yet popular conceptions of Indigeneity (in the academy and otherwise) have constructed slavery and African peoples' conquest-based removal from their Indigenous lands as disqualifiers for being counted as Indigenous peoples despite the fact that this occurred through imperialist and colonizing forces. And this disqualification occurs primarily because African Indigenous experiences under colonialism are generally not understood to be the definitive experience of Indigeneity. I argue that while undoubtedly tied to a distinct ancestral land or territory, Indigeneity can also indicate displaced communities' *relationship* with their homeland and must include an analysis of how imperial histories and social relations of power figure(d) in the displacement of Indigenous peoples from their ancestral territories. This analysis must consider their geographic proximity to, as well as their relationship with, these lands. By no means should my argument be taken as an attempt to deny the saliency and necessity of land with respect to providing self-determined sustenance for Indigenous peoples and their ways of knowing. Indeed, land theft and occupation are central to the implementation, execution, and perpetuation of colonialism and imperialist projects. Rather, what I propose here is a rearticulation of Indigeneity that permits the diverse histories and realities of distinct Indigenous groups to be seriously considered, so that the particular realities of diasporic Africans are included in this concept instead of being denied or subsumed under grand narratives that espouse the popular notions of Indigeneity. In this regard, we move away from exclusive or hegemonizing essentialist notions of Indigenous peoples that are narrowly predicated on obligatory residence on one's ancestral land, to a more complicated and inclusive approach that is reflective of the varied experiences of Indigenous peoples under colonialism and imperialism.

Being attuned to how imperialism and colonialism affect various Indigenous peoples *differently* also allows for a multifaceted conception of Indigenous identities that anchors but does not lock them into their ancestral land or territories. I do not conceive of being locked and anchored as synonymous. In my opinion, being anchored means *being grounded* with the promise of flexible movement, and this is not limited to solely a physical state of being. Interwoven in this concept are the psychic, emotional, and spiritual qualities of existence as well. Inversely, I conceive of the notion or sensation of being locked in as a position that precludes movement much in the same way certain definitions of Indigeneity have precluded the experiences of those who are Indigenous but have different, non-residential, or more transnational relationships and experiences with their ancestral lands *precisely because* their experiences with/in colonialism and imperialism have been different.

For the many diasporic Africans who have been physically and forcibly scattered, moved, and displaced from their ancestral lands, numerous emotional, spiritual, and psychic connections to various peoples and cultures of the African continent have nonetheless been retained through reassemblages of key elements of African culture such as language, clothing, food, music, dance, spirituality, etc. Given the reality of globalization as a hegemonic process imbued with political, economic, social, and religious inequalities that induce the continued movement of peoples across multiple spaces and national borders, it becomes futile to evoke hegemonically fixed or static notions of any form of identity or people. My project challenges us to think about the relationship between Indigeneity and land differently. Understanding that such relationships are highly complicated and differ for African peoples across time and space opens up the possibilities for developing more nuanced analyses of the multiple yet distinct ways that Indigenous peoples are positioned in relationship to their ancestral lands, and how colonialism and imperialism have figure(d) in changing these relationships. As a project that prioritizes African Indigenous and anticolonial perspectives in the face of dominant postmodern theorizing as the singular valid approach to "rigorously" theorizing identities, imagining Indigeneity differently becomes quite difficult when applied to African peoples who are often seen as "not Indigenous enough" or "not Indigenous at all." These tensions become heightened when the focus is diasporic because many would argue that being in a diasporic context renders (especially) African Indigeneity impossible or simply nonexistent. In this respect, I am keenly aware that I write against the academic grain and in the face of incredible opposition. However, my aim here is not so much to engage the debate on "the politics of identity" (Indigenous or otherwise) so much as it is to engage an exploration of African and, more specifically, Yoruba Indigenous knowledges and identities *on their own terms*.

That said, folded into this third layer of my conceptualization of Indigeneity is the necessity for a conceptualization of diaspora that is also inclusive and cognizant of difference. Rather than conceiving diaspora as the exclusive preserve of specific geospatial regions—a position that Stuart Hall (1997) takes—I prefer Earl Lewis's (1995) suggestion that we live in "a world of overlapping diasporas," which are "interconnected and demarcated by race, class, color, and other factors" (779).

Michel Laguerre (1998) and Aihwa Ong (1999) present a similar line of argument in their suggestion for "more flexible or diasporic notions of citizenship [in order] to probe the multiple belongings created in diaspora" (as cited in Braziel & Mannur, 2003: 6). Lewis's understanding helps me to think of the Yoruba communities here in Canada as a layer of diaspora that overlaps with others. What other layers precede and/or interact with that of the Yoruba in

Canada?[2] The first of these are the African diasporic layers, shared and related but different in that I am distinctly thinking about Africans who are the direct descendants of enslaved Africans. So there are those layers, and how that particular form of movement was traumatically enforced with no element of choice or autonomy for those who had been kidnapped away. My mind then veered toward the Indigenous peoples of Canada as another set of layers. As the First Peoples of this land, I then wondered, could they be confined to layers? I was not sure as to how to conceptualize or approach this. I recognize that my social location as a Yoruba woman raised in Canada since the age of three, on land that is not Indigenous to me, could not to be overlooked. I wondered: What are the implications of simultaneously being Indigenous and Western, on land that you are not Indigenous to? What does it mean to be a resident and citizen of a land that continues to be occupied by colonial settlers and where the Indigenous peoples of that land are still being oppressed and marginalized? Overall, my reconceptualization of land and diaspora as they relate to Indigenous peoples was to underscore the central themes of difference and multiplicity. George Dei and Alireza Asgharzadeh (2001) also insist on difference as a key principle and concept in anti-colonial theorizing:

> Oppression should be looked at as a site encompassing varieties of differences, categories, and identities that differentiate individuals and communities from one another and at the same time connects them together through the experience of being oppressed, marginalized and colonized. (316)

Hence, approaches to defining Indigenous must involve a recognition of difference where serious attention is given to the multiple and diverse experiences that Indigenous peoples have under imperialism and colonialism. In the same way that it is necessary to acknowledge that Indigenous identity is multiple, it is necessary to recognize that Indigenous people's relationship(s) to their ancestral land or territory is also multiple. This is a call for a conceptual shift where, instead of Indigeneity being presumptively predicated on compulsory physical residence on one's ancestral territory, this is replaced by the appreciation that notions of belonging and connection to one's Indigenous land(s) take multiple forms, and should therefore not be limited to simply one experience of how belonging is manifested. This conceptualization of Indigeneity allows for the varied histories and realities of distinct Indigenous groups to be seriously considered, so that a more multifaceted and inclusive approach is utilized to reflect the nuances and varied experiences of different Indigenous peoples under colonialism and imperialism.

Fourth Layer: Diasporic Indigeneity as Empowering Resistance

The fourth definitional layer of how I conceptualize Indigenous directly builds on the third to underscore the crucial theme of resistance and empowerment that Indigeneity holds. For this layer, I work with the idea that colonized Indigenous peoples can claim or evoke an Indigenous identity even when they are not on their ancestral land because such ways of knowing are flexibly embedded and embodied. However, before proceeding with further discussion, it is important that my position not be confused with endorsement of the appropriation of Indigenous identities in the way that individuals who come from dominant communities do. Linda Smith (1999) elaborates:

> [Indigenous] has been co-opted politically by the descendants of settlers who lay claim to an "indigenous" identity through their occupation and settlement of land over several generations or simply through being born in that place—though they tend not to show up at indigenous peoples' meetings nor form alliances that support the self-determination of the people whose forebears once occupied the land that they have "tamed" and upon which they have settled. Nor do they actively struggle as a society for the survival of indigenous languages, knowledges and cultures. Their linguistic and cultural homeland is somewhere else, their cultural loyalty is to some other place. Their power, their privilege, their history are all vested in their legacy as colonizers. (7)

While everyone is Indigenous to some place, sometime, somewhere (Ward Churchill, 2003b), a distinction must be made between constructions of Indigeneity that are interlaced with historical and contemporary legacies of conquest, colonial occupation, and White supremacy, and those that are grounded in struggle, resistance, and decolonization. The Indigeneity of the former is conveniently evoked by the dominant in certain contexts and shaped by power imbalances in the sense that such evocations do not disrupt their "innocence" (Razack & Fellows, 1998) as settlers. Settler identities such as this are heavily anchored in romantic amnesic constructions of themselves as benevolent founders. The Indigeneity that I speak of is one that is anchored in having been colonized on ancestral territory or a land mass different from the one being currently resided on. Especially for African diasporic peoples, this particular (re)configuration of Indigeneity works with the salient principle that Indigenous ways of knowing and understanding the universe are dynamic, and therefore flexibly embodied and embedded.

However, an Indigenous identity where one can claim citizenship on land that one is not Indigenous to raises important questions about power. As a

Canadian citizen, my Indigeneity is precariously built on the amnesic denial of Canada as stolen land. This is to acknowledge the paradox and complexity of being both Indigenous (Yoruba) and Canadian (Western) as a rather unsteady social location that does not conveniently fit popular notions of what it means to be Indigenous. Yet, I maintain that these realities do not make African peoples any less Indigenous. Rather, it is a *different form* of Indigeneity that inevitably highlights the third definitional layer, which emphasizes difference. I conceptualize this particular type of difference through something I term as a type of "diasporic Indigeneity." In essence, this is where one retains Indigeneity through cultural embeddedness and embodiment. I am remembering that constructions of Indigeneity vary, and are not singular, but diverse and dynamic. Therefore, including the Indigenous identities and knowledges of people who *carry* this (intergenerational) information with them in their physical bodies is legitimate, and considering how this operates in concert with their cultural histories and memories is worthy of study. This transpires despite the fact that they may not be residing on their Indigenous lands.

Thomas Heyd (1995) has argued that Indigenous knowledges are "embedded in distinctive social practices and cultural frameworks" (70) such as one's cultural memory (Dei, 2000). Some of the most profound examples of the retentions and reconfigurations of African Indigenous knowledges are found in the older African diasporic communities that were forcibly removed from the African continent during the Maafa, more popularly known as the European transatlantic slave trade.[3] Nevertheless, despite the flexible dynamism of Indigenous knowledges as "living knowledge systems that are continually responding [and adapting] to new phenomena and fresh insights" (Battiste, 2002: 12), the evocation of an Indigenous identity remains highly contested with continual charges of essentialism that equate the term with colonialism and imperialism—particularly in academic spaces. I echo the contrasting standpoint of scholars Isidore Okpewho, Ali Mazrui, and Carol Boyce-Davies who insist that memory of Africa and a sense of roots are, in fact, "political statements" and "psychological necessities" that serve(d) exiled Africans well, particularly when the conditions of oppression and colonization seemed intolerable (Okpewho, Davies, & Mazrui, 1999: xv). Similarly, George Dei and Alireza Asgharzadeh (2001) maintain that many aspects of Indigeneity act as "profound sites of empowerment, struggle and resistance against imposed hegemonies" (318). In "Essentialism, Memory and Resistance: Aboriginality and the Politics of Authenticity" Andrew Lattas (1993) presents one of the most thought-provoking positions on this issue. In agreeance with Okpewho et al. and Dei and Asgharzadeh, Lattas also posits that "cultural and political significance is always constituted through a past" where "a sense of continuity with the past might be a way of resisting assimilation" (246). Of principal significance is Lattas's response to charges of essentialism, which he establishes through a rearticulation of Indigenous identity *from an Indigenous worldsense standpoint:*

> An enormous amount of intellectual energy is currently directed at establishing Aboriginality as something that is invented through European involvement. What is often ignored is the sense of autonomy from the control of the "Other" conferred by images of the past and indeed the necessity to have an image of the past if one is to have a sense of ownership of oneself. Yet when Aborigines seek to give a mythological content to, or to reclaim, a primordial past for themselves then they are accused of essentialism and of participating in their own domination. Aboriginal culture here is set up to be demythologised and rationalized by the white intellectuals working in Aboriginal studies. It is to be stripped of its essentialising mythology and folklore and introduced to modern theoretical ideas which emphasize the contextual and relative nature of any identity. This is identity without content and without a primordial past; it is identity stripped to the bare logic of being simply a relation. The demand that Aborigines produce their popular consciousness along the lines of a social theory of identity is a request that they become conscious of themselves as purely relational identities; they are to be resisters without producing an essence for themselves. They are to situate themselves in opposition to Whites without fetishising themselves. They are to become a pure system of difference, an oppositional form that does not stabilize itself except through being a subversion of the other. There is no positivity and content in this form of Aboriginality, it is a relationship of opposition responding to the terms and agenda set yet again by white society. In effect, a white moral gaze refuses Aborigines an identity politics that is grounded in them taking up their bodies as an imaginary space. (247–248)

Lattas's brilliant discussion highlights the empowering possibilities of Indigenous knowledges as central decolonizing tools of resistance against domination and colonization. In placing Indigenous ways of knowing at the center, Lattas refuses the conventional Eurocentric yardsticks of academic identity construction and instead carves out a space to reconstruct an Indigenous identity on its own terms, that is, from an Indigenous perspective that affirms the past through descent, the body, and generational transmission of one's memories and ancestors. Lattas helps me to think about Indigeneity as a decolonizing terrain of intellectual engagement. He also demonstrates that essentialisms grounded in Indigenous constructions of the self cannot be solely and simply measured in relationship to Whites, but rather, Indigeneity also exists autonomously to serve important "cultural and political functions" (246) that are empowering, for example, being conscious of one's connection to one's past, one's land, and one's body as strategic ways of resisting colonialism and assimilation (245). This is not to say that relational readings and constructions of identity should be jettisoned or dismissed. Rather, what is being advocated here is the necessity to take seriously Indigenous constructions of identity on their own terms, and to be aware of the dangerous re-colonizing implications if these ways of knowing continue to be denied and delegitimized. In this sense, then, essentialisms that are "strategic" (Spivak, 1990) cannot be sloppily grouped with or subsumed under the oppressive and dangerously subjugating essentialisms of the dominant/colonizers. In effect, certain essentialisms are evoked differently because these formulations serve the function of strategic resistance, particularly in the necessary and eventual decolonizing shift toward healing from oppression. Such essentialisms are also therapeutic in that they provide physical, psychic, spiritual, and emotional ways of knowing and understanding oneself through a sense of community, groundedness, connection, and stability. In other words, certain essentialisms are empowering Indigenous constructions of identity that must be taken *on those terms*, and according to those ways of knowing the world, ways that are deeply anchored in their own unique Indigenous worldsense. Finally, Lattas's discussion reminded me that taking a staunch and absolutely anti-essentialist position where people are continually reduced to "pure systems of difference" (Lattas, 1993: 248) is a form of essentialism in and of itself because one remains fixed in that system. Indigeneity here is conceptualized both as an empowering resistant identity and *as part of a diverse range of identities*. The fact remains that there are myriad ways to understand our existence and identities in this world. And to exclude Indigenous identities, or slap them with charges that ring of some type of subtextual primitive essentialism, while engaging in a particular type of one's essentialism is to participate in exclusionary hierarchical politics that (intentionally or not) continue to deny Indigenous peoples their right to self-identify. In the end, such accusations are misguided and, in addition to reifying dominant Western knowledges, they fail to recognize the empowering and decolonizing possibilities of Indigenous knowledges and identities.

Fifth Layer: Avoiding Dichotomized Knowledge Construction

The fifth conceptual layer of my understanding of Indigeneity involves the premise that binaristic thinking around Indigenous and Western knowledges as clearly demarcated needs to be avoided (Dei, 2000; Purcell, 1998).[4] However, it is also important to recognize the current position of Indigenous knowledges and identities as "discredited" (Morrison, 1984) and subjugated, particularly in relation to dominant Western discourses about "truth" (Foucault as cited in Purcell, 1998: 260). To bring it back to the African context, the two monotheistic religions of Christianity and Islam have historically constructed African Indigenous spirituality—this includes that of the Yoruba—as inferior, uncivilized, and backward. Where Islam had relegated African Indigenous religions to *al-Jahilliyya*, the time of Barbarism, Christianity viewed it "as pure paganism" (Olupona, 1991: 1). Needless to say, these hegemonic constructions persist to date, as evidenced in the often ostracizing and antagonistically hostile attitude toward proponents of African Indigenous spirituality. The fact that many Africans have internalized, converted to, and consequently live out Christian or Muslim identities in allegiance to these colonizing religious traditions poses some interesting questions around how African Indigenous religions figure in liberation and decolonizing projects. Accordingly, an anticolonial theoretical framework becomes imperative to my argument because it engages a critique of the denigration and disparagement of Indigeneity, particularly since such pathologizing is carried out in the name of "modernity" (Dei & Asgharzadeh, 2001: 301). In short, the anticolonial discursive framework appreciates, and therefore takes the position, that Indigenous knowledges and identities carry crucial elements of empowerment, resistance, and the basic human right to simply be who one is. In this sense, justice and social change are attendant imperatives when working with/in this framework. Scholars Ladislaus M. Semali and Joe L. Kincheloe (1999) elaborate:

> A central tenet [is] our belief in the transformative power of indigenous knowledge, the ways that such knowledge can be used to foster empowerment and justice in a variety of cultural contexts. A key aspect of this transformative power involves the exploration of human consciousness, the nature of its production, and the process of its engagement with cultural difference. (15)

Sixth Layer: Anchored in African/ Black Feminist Theory

The sixth layer of my conceptualization of Indigeneity involves a synthesis of the salient Black feminist principle where the various forms of oppression in our world, such as patriarchy, White supremacy/racism, classism, ableism, and heterosexism are conceptualized as interlocking and mutually sustaining forces of dominance (Collins, 1990; hooks, 1984; Lorde, 1984). Against the traditionally masculinist grain of anticolonial theory, George Dei and Alireza Asgharzadeh (2001) echo Black feminist sentiments in their rearticulation of anticolonial discourse to assert the necessity that all forms of oppression be approached as interlocking, and therefore of equal importance in anticolonial theorizing. However, given such rearticulations, it should not be assumed that research that may focus more extensively on one or more sites of oppression is immediately doing so at the expense of others and therefore reproducing these same forms of oppression. Instead, attention must be paid to the manner in which this is done and ways in which certain forms of oppression and domination are salient in certain contexts, or the need for focus and emphasis on certain sites because they remain under-researched and scantily theorized. Dei and Asgharzadeh provide a lucid articulation of how this can be done in a strategic and anti-hegemonic manner:

> The anti-colonial thought forwards a notion of critical gaze which could be maintained on any single category such as race, class, or gender, at the same time can refrain from subduing or subordinating other categories and sites of oppression. Such a gaze is not concrete and fixed. It is fluid and transparent. It constantly sees and observes colonial relations of power and domination, shifts from one site onto the other, resists all of them, but maintains a relatively heavier presence on any chosen category in a strategic gesture to be more effective. (312–313)

As mentioned earlier, scholarship that premises contemporary Indigeneity from an African diasporic perspective remains quite scant and under-researched. My argument gives the concept of contemporary African diasporic Indigeneity the same consideration that other social identities and forms of domination are given when thinking through the interlocking nature of oppressions. Said another way, what I am suggesting here is that—as a social location and identity that one both celebrates, yet is systematically oppressed under through colonialism—African diasporic Indigeneity needs to be included as a category of analysis. Doing so could only enrich and nuance one's scholarship, but more importantly, doing so indicates a counterhegemonic political act in academic spaces. It is to say that one is aware of and contesting the (destructive) dominance of Western knowledges as normative. Finally, to do so (whether one identifies as Indigenous or not) would be taking a powerful alliance-based standpoint that signals to others the importance of speaking out against the continued oppression of Indigenous peoples, our knowledges, and lands.

Seventh Layer: Critical Interrogation of Indigenous Knowledges and Humility

The seventh and final conceptual layer of my understanding of Indigeneity is two-fold and builds on the sixth. The first portion entails an awareness that any critical analysis which seeks to understand the interlocking social relations of power in diasporic Yoruba identities requires us to put aside the seductive lull of romanticizing our Indigeneity. We must recognize that some Indigenous knowledge systems carry sites of disempowerment, which are often evoked against women and other cultural or ethnic minorities (Dei, 2000: 8).[5] These evocations tend to be erected in the name of tradition or under the banners of culture such as respect for one's elders, and are used as justifications for inequitable social relations. With a conscientious awareness of how colonialism figures, these sites of oppression also need to be critically engaged and examined, not disregarded.

The second portion of this final layer emphasizes the importance of working with a philosophy of humility, acknowledging the incompleteness of knowledge. This means embracing the power of not knowing (Dei, 2000). In my opinion, this is one of the most important lessons that scholars in the academy and Western world can learn from African and other Indigenous knowledge systems. In a space where one's worth and status is judged on an over-privileging of the primacy of the mind and "knowing it all," opening oneself up to the humbling position of viewing oneself as a perpetual learner rather than an eternal expert can be a transformative tool for many in the academy and beyond. As a core element of Indigenous knowledge systems, humility challenges us to *think and feel* outside our individual selves by engaging circle-centered epistemologies that focus on our interconnectedness with each other. The hope is that this will assist us in developing a critical reflexivity and sustaining deeper critical understandings of how these connections among one another reinscribe our position(s) in this world.

Ultimately, anticolonial theory is central to African diasporic identities because it privileges Indigeneity as an empowering and crucially significant standpoint from which to understand the world. Further, this theoretical framework produces scholarship that addresses colonialism, imperialism, and other inequitable social relations of power to allow for a more historicized, contextualized, and nuanced understanding of how specifically Yoruba Indigenous knowledges and identities are constructed, and how they might be reinscribed as empowering decolonizing tools.

African/Black Feminist Theory

My argument for diasporic Indigeneity is also anchored in African and Black feminist theoretical frameworks, both of which complement and contribute to anticolonial theory. These frameworks explicate and further nuance the complexity of African life and experience on the African continent and its diaspora. This is specifically done through the tenet of commitment to dismantling forms of oppression such as patriarchy, White supremacy, classism, and heterosexism. These discourses emerged largely from the unique positions Black women hold, where, on the one hand, we face the racism and White supremacy of White feminists, while on the other we face the (often internalized) sexism and patriarchy of African/Black men, all within the larger contexts of capitalism and Eurocentric societies (Amadiume, 1997; Collins, 1990; Combahee River Collective, 1974; Davis, 1981; hooks, 1981; Mama, 1995).

Whereas African and Black feminisms do have distinct histories in terms of their academic development,[6] it is important to point out that they also heavily overlap and dialectically inform each other.[7] As Black feminist contributions to transnational feminist theory continue to gain a foothold in academic circles, the emphasis on critically examining as well as crossing national, economic, political, and cultural borders has grown. Consequently, this has induced a shift toward focusing on how the local affects the global and vice versa. This shift has increasingly lent itself to shining a spotlight on the impact of inequitable social relations in African women's lives everywhere. Black/African feminisms are indispensable to my argument because they provide frames of analysis that allow me to critically interrogate and expose the entanglements of masculinist, patriarchal, and White supremacist knowledge systems that deny and oppress the voices, contributions, and experiences of African women. These feminisms are powerful because they place Black women at the center of analysis (Amadiume, 1987; Collins, 1990; hooks, 1981) while simultaneously employing the concept of intersectionality; where categories of identity, oppression, and analysis such as race, gender, class, sexuality, and ability are conceptualized as inextricably interdependent, mutually sustaining, and of equal importance (Collins, 1990; Combahee River Collective, 1974; hooks, 1984; Lorde, 1984). Coined by African American critical race scholar Kimberle Crenshaw, "intersectionality" enables me to think through the complex experiences of Yoruba migrants without hierarchically prioritizing either gender- or race-specific oppressions (Amadiume, 1997; Boyce-Davies, 1994; Crenshaw, 1991), while eclipsing others.

In contradistinction to White feminist or Black male-centered articulations of oppression (which position African women's experiences and knowledges at the lower ends of such hierarchies), Black female presence and voice is centered through the concept of intersectionality. This idea becomes crucially imperative because it offers a framework through which to situate the diverse realities, experiences, and knowledges of Black/African women as central, both historically and contemporarily.

A melding of African feminist theory and anticolonial scholarship is also of necessity for my argument given that it considers Indigenous knowledges as crucial entry points and sites of resistance and empowerment against Euro colonial oppression. African/Black feminist theory complements, overlaps, and builds on anticolonial frameworks because it allows for more nuanced attention to be paid to the entanglements of gender and power, and how they figure(d) in colonizing and imperialist systems of oppression. However, while there is a proliferation of African/Black feminist literature that explores the question, meaning, and significance of gender as a primary social category, my entry point in this book is African/Black feminist scholarship that draws on and engages the category of gender from Indigenous perspectives. These feminisms are a necessary conceptual tool here because they critically interrogate spaces of inequity within Indigenous knowledge systems. At the same time they recognize that these spaces of disempowerment are challenged and critically interrogated *in the promise of transformation and hope for equitable social change*. In this regard, African feminist frameworks that embrace Indigenous knowledges appreciate that these spaces of disempowerment do not then become all-encompassing representations of the whole network of Indigenous knowledge systems—as they are often framed in colonialist discourses. Ultimately, Black feminisms that utilize and are anchored in Indigenous-centered philosophies understand that Indigenous ways of knowing are not singular monolithic epistemologies, and that the impact of colonialism and imperialism on these knowledges must be given critical attention. This is especially pertinent given the reality that these systems continue to oppressively affect our choices, voices, and overall lives. This is not to deny our agency and forms of resistance. Rather, it is to highlight the insidious nature of the legacy of colonialism, how it often obscures our access to knowledges—particularly Indigenous ones—and the assorted fields of empowerment that these knowledges offer.

There is a need for a shift toward the fact that notions of Indigeneity are diverse and can also be diasporic. Indigeneity therefore needs to be taken up so that it is not imagined as conventionally singular or hinged on close proximity to, and/or residence on one's ancestral land/territories. To do so would in effect mean the exclusion of the unique realities of Africans in the diaspora.

In the end, the concept "Indigenous" needs to be imagined differently and more flexibly where diasporic Africans are concerned so that these distinctive social positionings can be theorized in a manner where the particularities of African diasporic experience is given the value and credence it deserves. For Africans in the diaspora, diasporic Indigeneity is a fact of our Blackness.

Notes

1. Fama's Ede Awo: Orisa Yoruba Dictionary, p. 97.
2. I am aware that this particular layer I speak of is not the first Yoruba diaspora in "Canada." I acknowledge that many other Yoruba peoples who were enslaved preceded this; however, for the purposes of this research project I focus on the most recent Yoruba speaking communities who "voluntarily" migrated from the African continent in the late 1960s to the early 1970s and on.
3. See Patricia Jones-Jackson, *When Roots Die* (1987); Melville J. Herskovits, *The Myth of the Negro Past* (1941); Marion Kraft, *The African Continuum and Contemporary African American Writers: Their Literary Presence and Ancestral Past* (1995); Maureen Warner-Lewis, *Guinea's Other Suns* (1991), *Trinidad Yoruba* (1997), and *Central Africa in the Caribbean* (2003); Robert Farris Thompson, *Flash of the Spirit* (1984); Edward Kamau Brathwaite, *History of the Voice* (1984); John W. Pulis (ed.), *Religion, Diaspora and Cultural Identity: A Reader in the Anglophone Caribbean* (1999); Albert J. Raboteau, *Slave religion: The "Invisible Institution" in the Antebellum South* (2004); Sterling Stuckey, *Slave Culture* (1987); Catherine A. John, *Clear Word and Third Sight: Folk Groundings and Diasporic Consciousness in African-Caribbean Writing* (2003); Jason R. Young, *Rituals of Resistance: African Atlantic Religion in Kongo and the Lowcountry South in the Era of Slavery* (2007) for excellent studies of how Indigenous African culture is reassembled yet retained and continued in the African diaspora.
4. This is important primarily because Indigenous knowledges are often co-opted, appropriated, and subsumed under "Western" knowledges, without being acknowledged in their own right.
5. For example, while not the focus of my study, I ask, how does homophobia, or the silence and denial of same-gender relationships in African Indigenous knowledge systems, limit itself as a source of empowerment and decolonization? As I make this assumption around the denial of homosexuality in African traditional societies, I feel the urge to unsay it because, yes, although Africans do share many cultural and spiritual values, epistemologies, and histories, it is also vastly important to not homogenize African peoples, or see the continent as an undifferentiated monolith. The differences and specificities of distinct African societies needs to be seriously considered. For example, Malidoma Some has discussed the key role spiritual gatekeepers—who are often gay men—play among the

Dagara people of Burkina Faso. He is also currently writing a book on gays and lesbians as spiritual gatekeepers in traditional Dagara society. See his website, www.malidoma.org, for more information. Also see Boris de Rachewiltz, *Black Eros: Sexual Customs of Africa from Prehistory to the Present Day* (1964). Despite the heavily anthropological construction of African societies, the pictures and images in this text give some insight into the diverse ways different African societies value, manage, respond to, and deal with sexuality and the dangers in assuming that the markers of sexuality in the West can easily be transferred to African societies. According to this text, homosexuality is accepted in some societies and severely punished or seen as unnatural in others. Additionally, the often metaphorical, indirect, and proverbial salience of African languages and oral traditions must be seriously engaged where analysis and discussion of sexuality in African contexts is raised.

6. On the ancestral shoulders of the long history of Black enfranchisement and Black abolitionism, the academic concept of Black feminism emerged largely from the historically activist-based and grassroots experiences of Black women in the United States. See Zora Neale Hurston, *Their Eyes Were Watching God* (1990) and *Folklore, Memoirs and other writings* (1995); Gloria T. Hull et al., *All the Women Are White, All the Blacks Are Men, But Some of Us Are Brave: Black Women's Studies* (1982); bell hooks, *Ain't I a Woman: Black Women and Feminism* (1981); Audre Lorde, *Sister Outsider* (2007); Angela Davis, *Women, Race and Class* (1981); Patricia Hill Collins, *Black Feminist Thought* (1990) for in-depth discussion and theorizing of Black feminism, as an important site for critical analysis of Black female experience within oppressive social relations of power. This is not to deny the long-standing grassroots activism and resistance of Black women on the continent, the Caribbean, Europe, and North and South America, which occurred simultaneously and alongside their African American sisters; rather, it is simply to say that the emergence of academic and textually based Black feminist discourses was initially conceived in primarily U.S. contexts.

   Similarly, see Ifi Amadiume, *Male Daughters, Female Husbands* (1987), *Reinventing Africa: Matriarchy, Religion and Culture* (1997), and *Daughters of the Goddess, Daughters of Imperialism* (2000); Amina Mama, *Beyond the Masks; Race, Gender, Subjectivity* (1995); and Patricia McFadden, *Gender in Southern Africa: A Gendered Perspective* (1998) for academic texts that explicate and theorize African feminism. To contribute to the dearth of written literature placing Black women as subjects at the center, these texts emerged largely from the historical and contemporary experiences of continental African women. Also, in light of the significance of narrative as central to African Indigenous culture and ways of knowing, many other African feminist intellectuals have theorized and voiced African women's experiences through narrative or storytelling. See works by Ama Ata Aidoo, Chimamanda Ngozi Adichie, Buchi Emecheta, Tsitsi Dangaremba, Mariama Ba, Flora Nwapa, Yvonne Vera, and Bessie Head (to name a few).

7. Despite contentious debates over the distinct specificities of the terms "Black" and "African," they do overlap, are highly interconnected, and therefore cannot be clearly demarcated as separate. Hence, I use these terms interchangeably as a political signifier in specific reference to the various and multiple Indigenous peoples of the African continent—to indicate both those who were stolen away through the horrific European transatlantic slave trade and those who remained on the African continent and were forced to undergo

the atrocious traumas of colonialism and imperialism. In the spirit of unity and community, I also use these terms interchangeably as a counterhegemonic and political identity to critically underscore the unique positionings and social locations of Indigenous Africans in Africa, Europe, the Caribbean, North and South America, and Asia. This is not to mobilize a conflated conception of "Africans" and "Blacks," where we are homogenized as one large monolith that exists absent of difference; rather, my aim is to discuss and theorize the complex nuances of Indigenous African life—wherever that may be—through usage of these terms as the larger politically unifying milieu under which Black/African people can be named and identified.

· 3 ·

IN DIALOGUE WITH THE SOULS OF YORUBA FOLK

Engaging a Yoruba Worldsense—Overtly Christian, Covertly Yoruba

> Sometimes we drug ourselves with dreams of new ideas. The head will save us. The brain alone will set us free. But there are no new ideas still waiting in the wings to save us … There are only old and forgotten ones, new combinations, extrapolations and recognitions within ourselves—along with the renewed courage to try them out. And we must constantly encourage ourselves and each other to attempt the heretical actions that our dreams imply, and so many of our old ideas disparage.
>
> —Audre Lorde, 2007: 38–39

The Souls of Yoruba Folk is based on narratives gleaned from face-to-face interviews with 14 Yoruba elders and their children who are young adults. All Yoruba community members interviewed were based in Toronto, Ontario, Canada. I used the interview method because it allowed me to center and retain orality as a fundamental component of Indigenous culture and ways of knowing where stories, proverbs, songs, and other forms of narrative experience can be shared by participants. Orally based interviews also allowed me to hear Yoruba community members' voices, thereby encouraging their agency and power in terms of determining how they would engage the dialogue. This qualitative approach was preferred because it acknowledges subjects as involved in the knowledge-production process, sharing richly textured and nuanced experiential information that a positivist approach could not accomplish. In this sense, Yoruba folk were understood and engaged as agents of knowledge (Collins, 1990) rather than as "sources of mere data" (Rosenberg, 2000: xvi).

I also felt a small number of participants was more conducive for a study that involves spirituality because it is a subject that is likely to be richer and flourish in the personal setting of a face-to-face interview, where privacy and confidentiality are respected and protected.

Moreover, the interview method was also used because of its promise of dialogue, which has "roots in an African based oral tradition" (Collins, 1990: 212), and because, as bell hooks (1989) reminds us, dialogue is a humanizing speech between two subjects that challenges and resists domination (131). I felt that using interviews as my primary method would allow for the Indigenous African tradition of orality to be continued, while remaining anchored in its consequent promise of resistance.

Finally, oral interviews as methodology were used because it fulfilled two of the three learning objectives for this book outlined above, namely, that these interviews be the conduit to opening up a space toward engaging in critical and more affirming dialogue about Yoruba Indigenous knowledges and identities, in contrast to the ways that Yoruba peoples and our knowledges have historically been rendered invisible and/or bound up within racist colonial constructions. Patricia Hill Collins (1990) elaborates on the importance of dialogue as a type of connectedness that is vital to affirming one's knowledges:

> A primary epistemological assumption underlying the use of dialogue in assessing knowledge claims is that connectedness rather than separation is an essential component of the knowledge validation process. This belief in connectedness and the use of dialogue as one of its criteria for methodological adequacy has Afrocentric roots. In contrast to Western, either/or dichotomous thought, the traditional African worldview is holistic and seeks harmony. (212)

In utilizing the face-to-face interview as method, the disconnection, silences, and oppressive secrecy that African Indigenous knowledges and practices overwhelmingly exist in slowly started shattering and Yoruba elders and community members were brought into affirming public spaces where their knowledges could be recast, rewritten, and revitalized as sacred. This was accomplished through critical, respectful, and engaged dialogue.

All of the interviews involved open-ended questions designed to stimulate dialogue and were carried out over a four-month period, from December 2006 to March 2007. I identified the community members interviewed through my volunteer work in the Yoruba community (i.e., at various Yoruba community organizations). During the initial stages of recruitment, I explained the project to all potential participants, gave all prospective interviewees time to get back in touch with me by leaving a phone number and email address so that they could contact me if, or when, they were interested in pursuing the project further. All of the folk featured in this book either had a parent or adult child who had also agreed to participate in the project. This entailed speaking to both the parents and their children on separate occasions to ensure that one did not pressure the other to participate. If both the parent and the child(ren) did not agree to be involved, then neither was interviewed. All Yoruba elders and youth were asked to sign a consent form. This process commenced only after having received approval and clearance from the university's research and ethics protocol.

Overview of Participants: Yoruba Elders and Youth

The people featured in *The Souls of Yoruba Folk* comprise a diverse group of Yoruba elders and at least one of their adult children. It was essential that a parent and at least one of their children participate because of the focus on intergenerational knowledge in this book. While 16 people agreed to participate, in the end a total of 14 elders and community members from five different families were interviewed. Regrettably, two participants—a single mother and her son—withdrew from the project due to personal circumstances that made them unavailable. Yoruba community members were previously asked to complete a biographical profile, which has been compiled into table format (Table 3.1) indicating their name (to retain their privacy and confidentiality, all participants were given pseudonyms for both their first and last names), age, gender, place of birth, years in Canada, educational background, marital status, and number of children. These categories were used because they helped to provide a more personal dimension to the data, particularly in terms of how it shaped my understanding and analysis of how Yoruba community members understood their Indigenous identities. The biographical profile was also used to document and illustrate the diversity in age, class, education, and gender to produce a richer perspective on how these factors may influence and shape folks' conceptions of their Yoruba Indigenous culture and identity. Of the 14 people who participated, eight were elders (the parents) who ranged from 45–59 years of age, while their children (the youth) were a total of six ranged from 18–26 years of age. Overall, I selected Yoruba elders who had raised their children in Canada for at least 10 years to gain a deeper understanding of the difficulties and fluctuations they may have faced while becoming accustomed to life in Canada. I also implemented a "ten years in Canada" criterion to gain a better understanding of how residence in Canada has influenced and shaped community members' views, comprehension, and practices of their Indigenous identities.

Ultimately, this group of Yoruba people generates a new perspective and approach to the study of Yoruba diasporic culture. Accordingly, *The Souls of Yoruba Folk* helps to fill a gap in research on the construction and lived practice of Yoruba Indigenous identities in dominant Euro-Canadian contexts.

The following provides a brief biographical description of each person, whom I discuss individually and within their larger familial structure. I first discuss the Yoruba elder participants and then the youth (their adult children). A condensed version of this biographical information can be found in Table 3.1.

Table 3.1 Participants' Biographical and Demographical Profile.

Participants	Age	Gender	Place of Birth	Years in Canada	Education	Marital Status	Number of Children
Sade Oriola	45	Female	Nigeria	16	Bachelor	Divorced	2
Mrs. Olusanmi	50	Female	Nigeria	26	College	Separated	2
Mrs. Awoniyi	58	Female	Nigeria	25	Bachelor	Married	4
Mr. Awoniyi	57	Male	Nigeria	25	Bachelor	Married	4
Mr. Fayemi	59	Male	Nigeria	32	College	Married	5
Mrs. Fayemi	46	Female	Nigeria	19	Bachelor	Married	3
Mrs. Oladiran	55	Female	Nigeria	25	High School	Married	4
Mr. Oladiran	57	Female	Nigeria	26	College	Married	4
Dele Oriola	23	Male	Nigeria	10	Bachelor	Single	0
Niyi Olusanmi	25	Male	Canada	24	High School	Single	0
Bisi Awoniyi	23	Female	Canada	15	Bachelor	Single	0
Seun Fayemi	18	Male	Canada	17	High School	Single	0
Yinka Oladiran	26	Female	Nigeria	23	Bachelor	Married	0
Tunmi Oladiran	23	Female	Nigeria	20	College	Single	0

Beginning each of these introductory biographical descriptions is a representative quote that I believe typifies the soul of each Yoruba person, together with a brief discussion of each participant's community affiliation, spiritual practice, and connection to Yoruba worldsense. In keeping with the title of this book, I hope the reader will appreciate these introductions as slightly more personal narratives written with the attempt to have you meet and acquire a more intimate sense of the people who make up *The Souls of Yoruba Folk*.

Ms. Sade Oriola (Elder)

> Everybody worships in different ways. I think it has to do with different denominations in the church. But I know it's still the same, when they talk of Ifa [Indigenous Yoruba spirituality] it still has to do with God … It's only worship in [a] different way … You know, before Christianity or Islam came to Nigeria, into my country, they have deities: that is, the Yoruba gods like Oduduwa, God of Iron and all those deities are messengers of God. And some people still practice it. Like in my family we have three!! Three in one! We have Christianity, Muslims and the pagans, the people who worship idols … But I was born into Christianity … But when they do festivals we still attend. So we're still … one way or the other, we're still connected to that type of religion.
>
> For me, yes I am in Canada, and what we call *ilu oyinbo* [the Yoruba reference to white people's land], but as Africans, we see the world in a different way. Um, we have ah, those that we can see, you know like us humans or people. But we also have those we can't see, at least not with our eyes. And it is all here too, it is all here; a part of the world no matter where you come from or go. This is what we Yoruba believe—and it is all owned and created by God, the highest power. This is very common knowledge for us Yoruba, and most Africans I know anyway. My sons, they know this too, even though we are here they know this. Like, no matter what, for me, they [our children] should know about the culture. It's very important because if you don't—like if they don't know about their culture, they will get lost. And everybody have their own culture. No matter what, even if you integrate into the Western world you still have to know your background, you still have to know your culture, and you still have to know that the way we do things, if it's different it's okay if we know God in our own way. So, it's important for me, but I don't know about others because some of our kids here don't know. But then, I know that definitely if you let them know about it, like if you teach them then they will learn and they will know.[1]

Ms. Sade Oriola, who identifies herself as Yoruba first, Nigerian second, and then Canadian, was born in Nigeria in 1962. A divorced single mother, Sade reminisced about her life in Nigeria with fondness and specifically spoke of Nigeria as a place where people were there for each other. She shared that with the community presence, she was able to have support in raising her children. However due to marital challenges and political problems Sade left Nigeria for Canada at 29. For her, Canada was an attractive place to live because she had heard of its multiculturalism and that it was a family-friendly country. To Sade this translated into meaning that her hopes of immediately sending for her two sons—who were still in Nigeria—would soon be realized. However, it would be another eight years before her sons could join her in Canada.

Sade was heavily involved as a volunteer and staunch advocate of Toronto's Yoruba Community Association (YCA). Ms. Oriola had lived in Canada for 16 years and during that time, she had earned undergraduate degrees in sociology and social work to upgrade herself. She spoke of deferring her desire to go to graduate school so that her children could attend university. The interview with Ms. Oriola was quite enjoyable and lengthy because of her great skill as an avid storyteller, a dancer, and an actor in the Yoruba and African Canadian communities. She advocates education and cultural knowledge with a special emphasis on "respect for elders" as the most important elements of Yoruba culture the next generation could possess. Sade and the youngest of her two sons, Dele, were recruited through the Yoruba Community Association and agreed to participate in this project.

As indicated in the above representative quote, while Sade identified as a Christian, her spiritual practice is fluid and progressive in that she acknowledges different forms of spiritual worship as different yet equal. Included in this is Sade's open embracement of her Indigenous Yoruba spirituality. This is evidenced in her discussion of attendance at traditional Indigenous festivals and acknowledgment of family members who practiced their Indigenous Yoruba faith. Moreover, the second portion of the representative quote highlights Sade's critical consciousness of Yoruba worldsense as multisensed and more complex than overreliance on the visual/tangible in the way that dominant Eurocentric worldview functions. Of significance is that Sade embraces this knowledge as divinely given and understands that Indigenous precolonial ways of knowing the world were valid and created by God. As well, in emphasizing the importance of using Indigenous ways of knowing to navigate living in the barren land of *ilu oyinbo* (white people's country), Sade is astutely aware of being strategic to survive in a place that is invested in one's oppression. Finally, in this powerful quote Sade calls attention to the critical importance of passing these strategies and Indigenous ways of knowing on to her children, Yoruba youth, and the generations to come as the ultimate approach not only for survival but especially as a source of self-knowledge and self-love to conjure strength in a foreign land that does not nurture this. In this compelling way, Sade's critical awareness of and strong connection to the Yoruba worldsense is undeniable.

Mrs. Olusanmi (Elder)

> We have so many religions in Yorubaland. We have some people who are Christian, who don't believe in Christ … who believe in Native doing. Those people have their own power too and with their own belief they can do or undo. They can change day into night and they can change night into day … I respect it but I don't believe in it. I don't want to get myself involve in it because where I come from, in my own home we were brought up in a Christian way. So being a Christian I don't believe in that and I don't want to get myself involved in it but as a Yoruba person I respect it.[2]

Born in Nigeria in 1957, Mrs. Olusanmi identified herself as Yoruba, African, and Black. She has lived in Canada for more than 26 years and came to Canada with her now estranged husband, Mr. Olusanmi. She holds a college diploma in health care but was not comfortable disclosing her current occupation or type of employment. However, she did share with me her dream of one day owning her own catering business. She expressed the need for the culture to be continued through "respect for elders" and dressing up in traditional Yoruba clothing as the most important cultural knowledge that she would like to see the next generation retain. Mrs. Olusanmi and her firstborn son, Niyi, were recruited through community referrals.

Similar to Sade, Mrs. Olusanmi's religious identification was as a Christian. She spoke of regular church attendance two to three times per week as important for her spiritual practice and again emphasized that the youth showing respect for one's elders especially during church attendance was paramount. Yet Mrs. Olusanmi also shows a clear awareness of Indigenous Yoruba spirituality and Yoruba worldsense in her pronouncement that its adherents are so powerful to the extent that "night can be turned into day" and vice versa. While stating that she was respectful of the Indigenous Yoruba faith, Mrs. Olusanmi was also clear in her wish to not be involved with this element of her culture and distanced herself from it as a result.

Mrs. Awoniyi (Elder)

> Yes, religion is just you going to the church … just adhering to what is being said or what you're reading or what is happening in the church, but spirituality is in you. You

> sit down, you just think of what you are internally not outwardly. Not what you can do, [or] what you can achieve, but what's in you! That's spiritual.[3]

Mrs. Awoniyi was born in Nigeria in 1949 and referred to herself as Nigerian Canadian. She, along with her husband and four children, has lived in Canada for more than 20 years. She holds a university degree in nursing and works at a hospital in the same profession. She stated with beaming pride that her children were fluent in Yoruba due to their decision to go back to Nigeria for six years. She recounts that this was during the children's formative years. Mrs. Awoniyi stated that of most importance to her was her children's happiness and that they respect their elders.

Interestingly, even as Mrs. Awoniyi identified as a Christian who attended church on a weekly basis, it was evident that the definitive marker of her spiritual practice happened outside the confines of church while she was alone. What Mrs. Awoniyi described was the practice of being in tune with her *interior* self, similar to the spiritual practice of meditation that transcends religious doctrine. Her emphasis on this practice as the essence of spirituality is noteworthy in that the focus is on an element of one's self not tied to any particular religious system or code of belief, but rather is a practice that encourages communion with/in one's soul. This approach, together with her pride in her children's fluency in Yoruba, as well as the wish that they respect their elders, indicates Mrs. Awoniyi's strong awareness of seniority as significant to Yoruba worldsense.

Mr. Awoniyi (Elder)

> [T]here is a very thin membrane. You could pass or cross over, both, in whatever aspect of life. Spiritual I would say is a personal thing. Religion is a collective thing. Because if you're being spiritual, you don't tell anyone, you just know it's a feeling that you have. But if you say that you're religious, it is because there is a ritual you perform that everybody sees and then they associate and say, "Oh that guy must be religious because he goes to church every Sunday." … But to be spiritual, it's a very, very, very personal thing that is just between you and God.[4]

Born in Nigeria in 1950, Mr. Awoniyi described himself as Nigerian and Canadian. He has lived in Canada for more than 25 years, and despite having been recruited to Canada with a degree in mechanical engineering, he has been employed as a customer service worker at a home improvement retailer for the past 10 years. He describes "integrity" and being helpful to their community as the greatest wishes he has for his children because, for him, these qualities transcend financial difficulties and would give them a sense of self.

Similar to his partner Mrs. Awoniyi, Mr. Awoniyi also identifies silent time with one's interior self as the substance of his spirituality and spiritual practice. However, Mr. Awoniyi also discussed the importance of church attendance within the context of community and collectivity as significant for one's spiritual practice. While admitting that he attended church, Mr. Awoniyi did not classify himself as Christian and instead said that "I am highly traditional! I have very, very deep cultural beliefs despite the fact that I am in the Western hemisphere of the world does not really change the values I have grown up with and which I grew up in … they are still quite ah, fresh and we tend to keep very close to it."[5] Mr. Awoniyi also began the interview/dialogue with a prayer of thanks to God for his children, partner, and family and repeatedly spoke of the importance of elders. Given his emphasis on community, having a strong connection to one's community and holding traditional values in high regard, Mr. Awoniyi also demonstrates his deep respect for and connection to Yoruba worldsense understandings of our world.

While a family of six comprising three girls and one boy, it was Mr. Awoniyi, his partner Mrs. Awoniyi, and their youngest daughter Bisi who were interviewed for this book. They were recruited through referrals from other members in the Yoruba-Toronto community.

Mr. Fayemi (Elder)

> I was told a story: when the Fulanis were coming, they wanted to invade the Yorubas and we pushed them back! By the power of our spiritual armor and rights you know. And if we had continued to associate our local religion with the Christian we would still be more powerful! Unfortunately we turned away what we have inside our house and we just bring in—it's like you have a hoe, the hoe that is working for you, that plants your plantain, your yam. Yet you throw it away because somebody brings you a shovel … we should have kept our own power. African power. I believe in African power and we have it.[6]

Mr. Fayemi describes himself as Yoruba and Canadian. He was born in Nigeria in 1948 and has lived in Canada for more than 32 years. He was recruited to Canada in the 1970s as an engineer, yet has spent the last 20 years working as a cab driver. Nevertheless, Mr. Fayemi still holds out hope that he will realize his dream of being an entrepreneur and owning his own business one day.

Additionally, his most heartfelt desire for his children extends to the larger Black community, where he wishes for the next generation(s) to "stay away from crime [to] be more successful and have high positions in business and government."

Mr. Fayemi identified as a Muslim but held that he did not go to Mosque or attend church with his wife and children, all of whom he said were Christian. He spoke passionately about Indigenous Yoruba spirituality and reminisced about attending many Indigenous festivals and ancestor masquerades in Nigeria where he would dance and sing along with the Native Yoruba worshippers. From his strong support of Indigenous spirituality as essential "spiritual armor" to collective African power and understandings of self, Mr. Awoniyi's notion of himself as a Yoruba man is consciously wedded to the larger African collective. Hence, his approach of emphasis on the self as connected to one's community and collective is in keeping with this as a core philosophy of Yoruba worldsense.

Mrs. Fayemi (Elder)

> I don't believe that you have to go to church to be spiritual. Your heart, your mind, whatever you do every day will show to people the type of person you are, the type of spirit you have. You can be going to church every day and the way you treat people doesn't reflect your spirituality. So it's the way or manner you treat people, or [how] you conduct your life that shows the type of spirit you have.[7]

Mrs. Fayemi was born in Nigeria in 1961 and referred to herself as African, Canadian, woman, and mother. She has lived in Canada for 20 years and holds a bachelor's degree in business administration; however, she is currently employed as a health care worker. She is also passionate about the Black community, the importance of education, and that her children have respect for their parents, as well as all elders in the community.

For Mrs. Fayemi, spiritual practice is located in one's integrity, how one behaves and how one treats others. Hence, although she spoke of going to church on a regular basis, for her, attendance was not enough as a true determinant of spiritual practice. This was evidenced in Mrs. Fayemi's discussion of observing people's behavior and treatment of others in church as a better way to know who someone really was. For her, this was a reflection of one's spirit. In this way, Mrs. Fayemi provides an important critique of religious identification, and specifically church attendance, as shallow measures of a person's spirit or soul. Combined with her passionate discussion of educating her children about racism, the importance of community and respect for elders suggests Mrs. Fayemi's awareness of and association with a Yoruba worldsense is strong.

Mr. and Mrs. Fayemi and their son Seun (the third son of five) were also recruited through referrals within the Yoruba community.

Mr. Oladiran (Elder)

> Spirituality, the way I would interpret it is, "where do you stand with God?" Religion can be anything; anybody can put anything together and say he's doing a religion. If you don't know what you are doing, people can use religion to vandalize, to destroy anything. If you want to talk about spirituality, look at where you stand as a spirit person, as a spirit being, the way God created you. Look at where you stand with God and religion, not so much.[8]

Born in Nigeria in 1950, Mr. Oladiran described himself as Nigerian Canadian. Mr. Oladiran is active in the Yoruba community and heavily involved with the Yoruba Community Association (YCA) in a leadership role. He spoke with immense pride in belonging to an association that could "support the Yoruba people and our culture and community in Canada." In his view, this was important because the next generation needs to know that their Yoruba community organization provides a space to keep the Yoruba culture alive. Mr. Oladiran holds a diploma in business management and retains belief in the dream of securing a job in this profession, despite the reality of having been a cab driver for more than 20 years.

Similar to Mr. and Mrs. Awoniyi, Mr. Oladiran advocates internal communion within one's self as essential to healthy spiritual practice. He shared that reflective examination of one's spirit or soul helps locate where one stands in relationship to God. As a community leader who advocates cultural continuity in the form of respect for one's elders and passing on the Yoruba language to the youth, Mr. Oladiran's connection to Yoruba culture and these elements of Yoruba worldsense are evident. Worth noting is the shared wish among Yoruba elders that their children have and show respect for their elders.[9]

Mr. and Mrs. Oladiran, along with their two daughters Yinka and Tunmi, were recruited through the Yoruba Community Association and all four members agreed to participate in this project. Their other children, two boys, chose not to participate.

Mrs. Oladiran (Elder)

> Spirituality is just power, power of God. Religion is just a name. A lot of people, when you see how they act, you ask them, are they going to church? [They say] "Oh I read my Bible" but it doesn't portray their life … But when you spiritually grow you don't have to ask before they see you and see the Glory of God upon you. Wherever you go, [there] should be a light. It's about the relationship you have with God and not what people see.[10]

Mrs. Oladiran was born in Nigeria in 1952 and also identified herself as Nigerian Canadian. An active member of Toronto's Yoruba Community Association, Mrs. Oladiran proudly espoused her husband's leadership role in the organization and spoke of her goal of having more youth involved. She holds a high school diploma and is employed as a cleaner in a health care facility. Like her husband, she has lived in Canada for more than 25 years and is adamant about the necessity of giving her children a better life through "prosperity and having the fear of God [instilled] in them."

Similar to her husband and Mr. and Mrs. Awoniyi, Mrs. Oladiran felt that religion was a narrow label that did not adequately reveal one's spirit or soul. Instead she felt that a person's behavior would show one's spirituality and whether or not one had "the glory of God" upon them. For Mrs. Awoniyi, her understanding of spirituality was heavily grounded in a Christian belief where church attendance was important for one's spiritual growth and, it seemed, central to her spiritual practice. She felt that reading one's Bible was not enough. However, Mrs. Oladiran also alluded to one's personal relationship with God as most valuable rather than what people are able to observe or see. Her emphasis on relationship and connection to God, rather than what's visible to the eye, is in keeping with Oyewumi's (1997) discussion of Yoruba worldsense. However, Mrs. Oladiran did take a strong position that in Nigeria, the time before Christianity was "a time of ignorance and not knowing God."[11] In her view, it was a time that Yorubas were not civilized and only "illiterates in the village practiced it."[12] For this reason Mrs. Oladiran's approach was quite in line with Eurocentric understandings of African spirituality as defunct, backward, and primitive. Accordingly, her understanding of Yoruba worldsense was quite narrow and static.

Dele Oriola (Youth)

> Religion I think of it as, um, organized. The more formal part of it; like the Catholic faith. Spirituality is just you yourself. Like it's more personal. I believe it's for you to look inside your own self.[13]

Twenty-three-year-old Dele was born in Nigeria and described himself as both Yoruba and Black. At the time of our interview, Dele was in the process of completing an undergraduate degree in the arts. Of particular interest is that his mother was displeased with his transfer from a pre-medicine program to one in fine arts, despite the fact that she herself is a talented artist. Dele expressed particular interest in pursuing a career in music and entertainment, and was especially fascinated with Nollywood, the Nigerian film industry, which is the third largest in the world after Bollywood and Hollywood. Like his mother, Dele feels that the Yoruba culture needs to be taught to the younger generations, with a particular emphasis on having respect for one's elders and speaking the Yoruba language. Dele shared his experiences of racism and staying out of trouble as a way to not contribute to stereotypes of Black people and especially young Black males.

Like many of the Yoruba elders interviewed for this book, Dele's understanding of spirituality was one that was personal and included reflection and being in tune with one's interior self as important. For Dele, introspection was an important part of his spiritual practice. Evidence of this can be found in Dele's soul-searching, which resulted in his decision to leave pre-medicine and follow his passion for the arts and entertainment, in his mother Sade's footsteps. Similar to his mother and other Yoruba elders, Dele was a staunch advocate of respecting one's elders and passing on the Yoruba language to younger generations. Given his position on these aspects of Yoruba Indigenous culture, Dele's relationship to these elements of Yoruba worldsense is compelling.

Niyi Olusamni (Youth)

> I like going to church because it helps me be close to God. I think church is important and I try to get my friends to come all the time but they are not really interested. My moms taught me this was important and I respect her and my elders. The part I like best is that it's a Nigerian church and it's important that I know my heritage so that when I have kids I can pass it to them and when I die they'll have something to continue.[14]

Born and raised in Canada, 25-year-old Niyi described himself as a Canadian of Nigerian descent. He held a high school diploma and was still not sure what type of career path he wanted to pursue at the time of our interview. Although single, Niyi cited settling down, getting married, and starting a family as his main priority and looked forward to moving out of his mother's home to start living on his own. Interestingly, Niyi also spoke of racism as a form of bullying he had experienced as a child and expressed that he did not want his children to have to experience it. He communicated that this would be a success for him. When I asked if he would elaborate on the racism he declined saying that he did not want to discuss it. From this part of our conversation it was clear that there was much pain for Niyi regarding the racism he endured and his way of dealing with it was in trying to forget.

It was also clear that his spiritual practice involved going to church because he felt this strengthened his relationship with God, and very much enjoyed being in the company of fellow Nigerians for Christian worship. Niyi was respectful toward his mother (Mrs. Olusanmi). He also felt strongly that his Nigerian heritage and the Yoruba cultural value of respecting one's elders were values he wanted to pass on to his family when he got married. When I asked for his thoughts on Indigenous Yoruba spirituality, he said that it was outdated and his mother had told him that "only illiterates back home practiced it."[15] However, Niyi did also admit that he did not know much about it. Given Niyi's perspective and lack of knowledge concerning his Indigenous spiritual culture, his access and connection to Yoruba worldsense is limited.

Bisi Awoniyi (Youth)

> Religion is a big part of my life. Huge. Huge. Not only because that's the way I grew up, because my family's quite religious and that's the way we were raised. But I guess you come to find your own spirituality because you go through a situation and I know prayer works … But some people will say, "I'm spiritual but I don't go to church." Well then who are you praying to?[16]

Twenty-three-year Bisi was born in Canada and identified herself as Nigerian. She is Mr. and Mrs. Awoniyi's daughter and the youngest in her family. At the time of the interview she was pursuing a degree in communications and expressed much interest in living in Nigeria. Despite speaking Yoruba fluently, she felt that she was "missing out on a lot of the culture" and wanted her future children to know it.

Bisi was outspoken about religion being central to her life and saw it as synonymous with spirituality. For her, prayer and going to church were essential to her spiritual practice. She also seemed to be doubtful of others who did not attend church and gave the impression that if one did not go to church one was somehow not worshipping in an appropriate manner. Additionally, when I asked Bisi about Indigenous Yoruba spirituality, she initially hesitated but then said that she thought of it as "bad medicine" that is primarily used to harm people. When I asked if she had any other knowledge of our Indigenous spirituality as Yoruba people she admitted that all she knew were stories her parents had told her and the Nollywood movies she had watched, which showed people going to medicine men to hurt other people. Bisi has a strong spiritual practice of prayer and church attendance that is informed by her Christian values, and her understanding of Yoruba worldsense is also heavily informed by this doctrine. In this way, her connection to Yoruba worldsense is parochial and shaped by Eurocentric Christian supremacy.

Seun Fayemi (Youth)

> If you're cool with your religion and you're like really close to the church and everything, I think you'll have a good spirituality. Like you'll be able to like have your calm times [and] handle your stressful times.[17]

As the youngest of the Yoruba youth interviewed for this book, 18-year-old Seun was born in Canada and identified himself as both African and Nigerian. He is Mr. and Mrs. Fayemi's son and was in his final year of high school at the time of our interview. He was also not sure that university was right for him. Like many youth at this stage of life, Seun did not yet know which career path he wanted to pursue, but did know that whatever he chose "it would be hard work."

Seun spoke of going to church on a weekly basis with his mother and two younger brothers. He associated good spirituality with Christianity and specifically with going to church. For him, this seemed to be the spiritual practice he used as an effective way to have more calm moments and manage stressful times in his life. When I asked Seun about his thoughts on traditional Yoruba spirituality he spoke of Nollywood movies doing "voodooism" and "seeing a lot of evil stuff where people are always trying to harm other people."[18] Similar to Bisi and Niyi, Seun's knowledge of Yoruba worldsense is limited and shaped by Nollywood movies, which construct Indigenous practices as the opposite of Christianity, modernity, and civilization. Given the dominance of Eurocentric discourse about Indigenous spirituality in Nollywood movies, and given that this is a popular yet hegemonic source of accessing information about Yoruba Indigenous practices, Seun's connection to and knowledge of Yoruba worldsense is restrictive. Of particular note is the emerging theme of Indigenous Yoruba spirituality as evil and harmful. Yet, given Seun's parents' respect for Yoruba spirituality and their strong connection to Yoruba worldsense, it is unclear why Seun has minimal knowledge of his Yoruba indigeneity.

Yinka Oladiran (Youth)

> In the way that I was raised, being a Christian I find is a powerful thing. And the Bible basically guides everything that I do and believe. That's the center of the whole world is my religion … Being a Pentecostal compared to Catholic or Anglican—'cause I've been an Anglican before—the way in which they worship is completely different. What I'm saying is Pentecostals—well I'm not gonna speak for everybody else but, the Pentecostals that I know have the Holy Spirit and that's what we work with is the holy spirit and it says so in the Bible. So I can't speak for someone that's Catholic, someone that's Baptist and someone that's Presbyterian but I feel because I'm Pentecostal, I have a sense of power because I have the holy spirit with me.
>
> If I think back to my Grandfather's generation, and even going further than that, they were Christians! Like not in the sense of being spiritual, or in the sense of going to herbalists and stuff. And when I say herbalists I mean like, um, voodoo doctors and stuff like that. No, he was a Christian. He didn't know how to read or write, so he didn't read the Bible. But he knew how to pray. And he knew that there was a God and he used to pray like that. Um … my sense of ah, people that go to spiritualists, or go to herbalists and stuff, those are people that … they're spiritual yeah. But they're not spiritual in a good sense … Most times people that go to those kinds of things are people that want to do harm to other people.[19]

Twenty-six-year-old Yinka Oladiran was born in Nigeria and came to Canada with her family when she was five years old. She is Mr. and Mrs. Oladiran's eldest daughter and identified herself as a Canadian. She holds an honors bachelor's degree in sociology and was employed as a court officer at the time of this interview. She was also recently married and expressed much excitement about her new life as a wife, citing that she especially looked forward to being a mother in the near future.

Yinka also expressed staunch pride in being a Christian from the Pentecostal denomination. She shared her love for her church due to its powerful praise and worship of Jesus Christ as central to her life. As spiritual practice she spoke of working with the Holy Spirit, the power of prayer, biblical engagement, and weekly attendance at her church. Yet, drawing from the above representative quote, Yinka is also quite invested in rather exclusive Christian lineage that denies any ancestral practice of, or participation in, Indigenous Yoruba spirituality. Comparable to other Yoruba youth in *The Souls of Yoruba Folk*, Yinka expressed that Indigenous spiritual practices of Yoruba people were spiritual "but not in a good sense" and characterized these practices as harmful. Yinka's position is one that is heavily informed by Eurocentric Christian discourse, which has constructed Yoruba Indigenous practices and beliefs as evil, dangerous, and malevolent. Interpretations such as these are similar to the doom and devil-ridden view of the world Frederic and Mary Ann Brussat (1996) discuss as the cause for significant blockage to spiritual literacy. In this way, Yinka's perception of and connection to Yoruba worldsense is obstructed.

Tunmi Oladiran

> Well there are a lot of people that are not religious but they're spiritual. They have a relationship with God and they don't really connect themselves to any certain religion … I guess when I was younger I really didn't have much of an appreciation for my religion. Um, but as I got older I could see like, you know what I mean, like I could see my parents' values and the things that they hold dear to them, like reading the Bible and praying. And I even sometimes see it in me whenever I'm doing stuff so it's like, you know what I mean. Like it's something that you see growing up and either you respect it or you don't. But I get a lot of answers when I'm confused about something. The Bible guides me and I never used to see it but now, it's really important, an important source of spiritual guidance, like you know what I mean?[20]

At 23 years old, Tunmi Oladiran is the youngest in the Oladiran family and describes herself as African Canadian. She is Yinka's younger sister, was also born in Nigeria, and came to Canada with her family at three years old. Tunmi holds a college diploma in social service work and expressed interest in obtaining a university degree in the same profession because she found her volunteer work at the Jane and Finch Community Centre to be "very fulfilling and rewarding in giving back to the community."[21]

Similar to Dele and many of the Yoruba elders interviewed, Tunmi understood spirituality to be individualized and informed by one's relationship with God. For her, it was not necessary that it be connected to any particular religious doctrine or belief system. Interestingly, while Tunmi did not explicitly identify or attach herself to any particular religion, she did speak of her parents' values and the Bible as a source of strength, guidance, and support, especially as she matured. However, when I asked about Yoruba Indigenous spirituality she admitted that she did not know much other than Nollywood movies, which often show it to be quite negative and evil. Again the theme of Indigenous spiritual practices as evil is reasserted. What is specifically disturbing is how limited many of these Yoruba youths' awareness of or connection to their Yoruba Indigeneity is, as well as it being profoundly anchored in colonial constructs.

Inevitably, the prevalence of Eurocentricity obscures the possibility of critical spiritual literacy for many Yoruba community members and therefore highlights the urgency of this conceptual tool namely because this is a lens that deconstructs Euro-Christian discourses as dominant ways of knowing. This is a critical approach that also examines how these colonial inequities erect barriers to reading Indigenous practices as sacred. The sacred is not confined to dominant religions but is embedded within and written into our everyday lives and experiences, and this includes Indigenous spirituality. What follows is a more nuanced mapping and application of a critical spiritual lens, to make evident the consequences of shouldering Eurocentric discourses as normative. It is a consequence that too many Yoruba folk bear.

On Spiritual Closets and Concealed Indigenous Identities

> "Christian by day; pagan by night." If you go to the cities of Africa and you talk to people about village spirituality, they say, "Oh, no. These are not things that we do anymore. These are things of the past." People will stick by this view until it becomes a matter of life and death. But at night, in the cities of Africa, go to the diviners. You will see people there, sitting, waiting in line to see the diviners, because they are no longer able to make sense of their life.
>
> —Sobonfu Some, 2003: 89

Eurocentric and colonialist constructions of Yoruba (and, to a larger extent, African) Indigenous spirituality so deeply pervade the moral, social, and institutional fabric of Western society that negotiated engagements with these core aspects of our identities (i.e., ritual, ceremony) are often forced to occur in the peripheries of our social worlds, and largely in secret. This secret existence that many Africans have been forced to live is what I refer to as a spiritually closeted life. While this concept is conventionally built on ideas that specifically pertain to sexual orientation or queer identity (Butler, 1990;

Johnson & Henderson, 2005; Lorde, 1982; Sedgwick, 1990;), my discussion of "the closet," secrecy, and concealed identities is anchored in a different ideological and epistemological space, that is, within the framework of Black feminism, African Indigenous spirituality, and the cosmological philosophies that inform these ways of knowing and being. I utilize the Black feminist concept of intersectionality, and the work of scholars such as Audre Lorde, Zora Neale Hurston, and Patricia Williams, to highlight the complexity of concealed Indigenous identities and how they can be informed by inequitable social relations of power such as colonialism, racism, sexism, and one's Indigeneity. This is not to deny that the origins of academic scholarship produced around the closet come from foundational queer theory. However, the analogy of the closet helps me to theorize and discuss another wounded existence that is also deeply marginalized and confined to asphyxiating social spaces, which I argue injure the soul/spirit.

I recognize the irony of using the Western notion of the closet to denote a complex Indigenous social reality that exists both with/in and outside hegemonic Western culture. However, given the historical and contemporary insidiousness of colonialism and imperialism in Western and non-Western colonized societies, contradiction is a messy fact and inevitable reality of social life. In other words, I do understand that the closet is a Western cultural and historical construct (Sedgwick, 1990) and it may therefore seem problematic to some that I use this term to symbolize a rather different social life. However, I argue that exactly the opposite is true and that this discussion represents a truth. My position is two-fold. First, when a people's everyday cultural and spiritual life is so deeply disrupted, disparaged, and reconstructed as inferior (at best, if not completely wiped out for some), aspects of this social life become relegated to the margins, while new hegemonies of life that are disguised as modern and progressive pervade the center to become normal. It is dominance, oppression, and hegemony that have created this reality of marginalization. To effectively theorize and discuss life in the peripheries, one must name the systems that created the inequities in the first place. The West teaches us who live in this part of the world to hide and push what is not valued or respected into the closet. Colonialism created this fragmentation and its attendant wounds, hence building on similar peripheral social realities that are endemic to Western society become ideal sites to begin critical anticolonial interrogation of inequity, and how it can be effectively theorized and resisted.

Echoing Audre Lorde above, the concept I introduce and discuss here is not a new idea. Rather, it is a new extrapolation of an old (Western) idea that, quite ironically, underscores the indispensable Black feminist concept of intersectionality. This is where the connections between different forms of oppression—be they spiritual or sexual—are illuminated. Hence, what on first glance may seem problematic I prefer to approach as a necessary contradiction that is part of the abstract messiness of colonial oppression and liberation: engaging the West to understand Indigenous life in hegemonic Western contexts. Black feminist theorizing—specifically the concept of intersectionality—has taught us that anticolonial political work is messy, contradictory, and not easy. Ultimately, the foundational tenet of all forms of systemic, institutional, and spiritual dominance is the violent and injurious insistence on singular notions of normalcy as the conventional standard that all must live up to and abide by. This sets the deleterious groundwork for confining spaces such as closets to be a part of the Western cultural landscape—albeit a hidden one—and how one becomes forced into them. Unsurprisingly, such hegemonies must then dismiss multiplicity and interconnection to keep these fixed and oppressive notions of sexual and/or spiritual normalcy intact. That said, I do find it necessary to reiterate that while the connection between the spiritual and sexual in terms of closeted peripheral identities is undeniable, my focus remains on theorizing the lived reality of what it means to be spiritually closeted from an African Indigenous perspective. My hope is that this work will act as a testimony to this complicated reality, while simultaneously providing a deeper understanding of the harmful effects of closeted Indigenous spiritualities that manifest in both the individual and collective psyche and souls of (too) many African peoples.

I arrived at the concept of spiritual closets through the voices and experiences of the 14 Yoruba folk I interviewed for this book. Their articulations of Indigenous Yoruba spirituality were pivotal to my developing and conceptualizing critical spiritual literacy. However, here, the idea of spiritual closets is introduced and theorized by answering the following questions: How are Yoruba Indigenous knowledges constructed by diasporic Africans of Yoruba descent? What are its manifestations, particularly in terms of daily life and experience in Canada? My focus is not so much on how Indigenous Yoruba spirituality is defined, but rather how Yoruba words, stories, ideas, and meanings are used and voiced by elders and youth in relation to their conceptions of Indigenous Yoruba spirituality. During my interviews with community members, words such as "juju," "idol," "idol worshipping," "pagan," "illiterates," and repeated discussions of "harm" and "evil" were often interspersed with furtive or veiled silences that had seemingly contradictory undercurrents, stories, and conversations surrounding them. Hence, I tease out and critically interrogate how community members conceptualize Indigenous Yoruba spirituality, and I use this analysis to arrive at *spiritual closets* as a term that provides a metaphoric image to convey the complicated reality of concealed African Indigenous spiritual identities. Ultimately, this provides a glimpse into the quality, substance, and consequence(s) of living a spiritually closeted life and the complexity of how Africans of Yoruba descent draw on and construct Yoruba Indigenous knowledges to navigate Eurocentric terrains of life in Canada.

Spiritual Closets: Whispers of a Concealed Spiritual Self

Yoruba people share a collective consciousness about the interconnectedness between the seen and unseen worlds, particularly where matters of spirit are concerned. This is largely the case because Yoruba understandings of the physical and metaphysical are anchored in what Oyewumi (1997) refers to as an Indigenous worldsense. The first question I asked Yoruba community members was to share their understanding of spirituality and how they conceptualized it. They overwhelmingly believed spirituality to be infused in daily life, both seen and unseen. While a number of people limited their conception of spirituality to Christianity, Ms. Oriola (or Sade) was one of the few who, in seeing spirituality as part of daily life, included Indigenous Yoruba beliefs. When asked about her understanding of spirituality, Sade (Ms. Oriola) responded that spirit was "everywhere and in all things, which [inevitably] go back to God." Sade also spoke of spirituality being in the church, as well as in Ifa (Indigenous Yoruba) practice. She stated:

> Everybody worships in different ways. I think it has to do with different denominations in the church. But I know it's still the same, when they talk of Ifa [Indigenous Yoruba spirituality] it still has to do with God … It's only worship in [a] different way … You know, before Christianity or Islam came to Nigeria, into my country, they have deities: that is, the Yoruba gods like Oduduwa, God of Iron and all those deities are messengers of God. And some people still practice it. Like in my family we have three!! Three in one! We have Christianity, Muslims and the pagans, the people who worship idols … But I was born into Christianity … But when they do festivals we still attend. So we're still … one way or the other, we're still connected to that type of religion.

For Sade, her individual and communal identity as a Yoruba woman is inseparable from Indigenous Yoruba cultural beliefs despite being raised as a Christian. For her, spirituality is regarded as something that precedes humanity and religious doctrines. Similarly, when Mr. Fayemi was asked to share his understanding of spirituality, he responded: "Well, for me, spirituality goes hand in hand with religion because the idea of being religious is attached to your spirit and spirituality is the core of your religion. Of course, you can also be spiritual and not be attached to any particular religion so spirit is bigger than religion because it is manmade." On the other hand, Bisi Awoniyi found my question to be directly and exclusively linked to Christianity, and spoke of other religions (such as Islam and Buddhism) as spirituality defunct and "not of God." Like Bisi, sisters Yinka and Tunmi Oladiran as well as Mrs. Olusanmi and her son Niyi also shared this narrow view of spirituality and regarded it as existing within the confines of Christian philosophy and doctrine.

I also asked community members to share their thoughts on the relationship between spirituality and religion. Given the contradictory and overlapping approaches to both terms, I wanted folks to reflect upon and explain whether one or both spirituality and religion were of importance to them, why, and how. Although the responses varied, there were a number of convergences and divergences that are worthy of note. Most noteworthy was the overwhelming consensus that spirituality (not religion) mattered more to them as individuals within the larger context of their Yoruba community. This is of significance because it opens up space for the possibility of critical dialogue about spirituality that both includes and extends beyond the hegemonic confines of dominant religion. The promises of such possibilities are empowering and not to be understated.

The second point of convergence among many of the Yoruba elders and youth was their identification of spirituality as distinct from religion because it was more personal, and involved one's private individual relationship between self and God. The following are some examples of responses to this question:

Mr. Awoniyi: Hmm. That is a very, very … there is a very thin membrane. You could pass or cross over, both, in whatever aspect of life. Spiritual I would say is a personal thing. Religion is a collective thing. Because if you're being spiritual, you don't tell anyone, you just know it's a feeling that you have. But if you say that you're religious, it is because there is a ritual you perform that everybody sees and then they associate and say, "Oh that guy must be religious because he goes to church every Sunday" or "there is a prayer group going on around him always, so he must be religious." But to be spiritual, it's a very, very, very personal thing that is just between you and God.

Mrs. Awoniyi: Yes, religion is just you going to the church … just adhering to what is being said or what you're reading or what is happening in the church, but spirituality is in you. You sit down, you just think of what you are internally not outwardly. Not what you can do, [or] what you can achieve, but what's in you! That's spiritual. And if you sit down and look around your surroundings and know what is happening, you will see many, many things which, inside you, you will now say, "oh this is different." But religion, yeah on Sundays, I just go and sing and clap and listen to the … it's a ritual thing. [As for] Religion, I go to church, I pray but spirituality wise, I sit down, I just source around things in me and if there are any changes I like to make, I try to do it.

Mr. Oladiran: Mmmm, okay, spirituality, the way I would interpret it is, "where do you stand with God?" Religion can be anything; anybody can put anything together and say he's doing a religion. If you don't know what you are doing, people can use religion to vandalize, to destroy anything. If you want to talk about spirituality, look at where you stand as a spirit person, as a spirit being, the way God created you. Look at where you stand with God and religion, not so much.

Mrs. Oladiran: Spirituality is just power, power of God. Religion is just a name. A lot of people, when you see how they act, you ask them, are they going to church? [They say] "Oh I read my Bible" but it doesn't portray their life … But when you spiritually grow, you don't have to ask before they see you and see the Glory of God upon you. Wherever you go, [there] should be a light. It's about the relationship you have with God and not what people see.

Mrs. Fayemi: I don't believe that you have to go to church to be spiritual. Your heart, your mind, whatever you do every day will show to people the type of person you are, the type of spirit you have. You can be going to church every day and the way you treat people doesn't reflect your spirituality. So it's the way or manner you treat people, or [how] you conduct your life that shows the type of spirit you have.

Dele: Religion I think of it as, um, organized. The more formal part of it; like the Catholic faith. Spirituality is just you yourself. Like it's more personal. I believe it's for you to look inside your own self.

The significant theme in these voices deserves critical attention: Despite community members' insistent identification as Christians, they nevertheless made a clear distinction between religion and spirituality that carried with it a number of subtleties suggesting a critique of religion, and the Christian church in particular, as expressed by Mr. and Mrs. Awoniyi, Mrs. Oladiran, and Mrs. Fayemi, all of whom constructed church attendance as a social routine that did not necessarily inform or reflect a person's spirit, character, or inner being. In this sense, church attendance can be read as a necessary external performance one engages in as a means of gaining respect in the community, as well as within larger social/institutional networks. Said another way, church attendance functions as a type of religious yardstick against which one's respectability and social approval is measured. Even the mere utterance or spoken claim of church attendance acts as a type of unwritten garment of approval that one dons, possibly to avoid arousing doubt or suspicion around one's commitment to their Christian faith and identity. Another nuance of the distinction between religion and spirituality was the shared identification of spirituality as a personal and private preserve that was not shared with others. Important to note once again is that this was especially not done within the confines of church. In this conceptualization, the church becomes a paradoxical space in which one deliberately conceals and does not expose the inner or most private spiritual self, despite the fact that one of the church's primary claims is that it provides religious and spiritual refuge from the secular world. For these individuals, spiritual identity is not anchored in the social, cultural, and physical structure of the church, but rather is identified as a private interior space inside the self. This is a space that has relationship and communion with a larger spiritual being, which was overwhelmingly identified as God by Yoruba community members.

Worthy of note are also the points of divergence, or differing responses given by Yoruba elder and youth with respect to their understandings of spirituality and how it is related (or not) to religion. Differently from other Yoruba community members, Sade, Mr. Awoniyi, and Mr. Fayemi maintained that being spiritual was not exclusive to Christianity, but also extended to other faiths—including Indigenous Yoruba spirituality—and the nonreligious. Again, this is of significance because folded into this understanding is a critical awareness of spirituality as larger than the dominant religion. These three individuals are actively resisting narrow hegemonic notions of spirituality and are instead choosing the empowering position that spirituality is diverse, layered, and can be engaged in multiple ways, forms, and contexts, as in the Yoruba worldsense.

The second significant point of divergence concerned how community members in this project located themselves as spiritual beings in response to my questions about spirituality and religion. While the vast majority of respondents situated their spiritual existence within Christian doctrine and philosophy, Ms. Oriola, Mr. and Mrs. Fayemi, and Mr. Awoniyi all openly declared that despite being raised Christian, there still remained with/in them an important connection to and identification with Indigenous Yoruba spirituality. Despite distancing themselves from open or explicit involvement in Indigenous Yoruba spirituality, these four individuals also spoke of dancing and singing with Native worshippers when public ceremonies or festivals were being held when they were in Nigeria. Such participation at first may seem contradictory to some. Yet this type of contradiction is one where Yoruba people are engaging in acts of resistance and insistence that celebratory participation in public Indigenous festivals or ceremonies is part of who they are as community members in the context of a larger Indigenous Yoruba collective. Inevitably, such festivals act as emancipatory spiritual spaces of identification for these folk.

Another way in which Sade Oriola, Mr. and Mrs. Awoniyi, and Mr. Fayemi identified and connected with Indigenous Yoruba spirituality was through speaking their Yoruba mother tongue. It is not a coincidence that language is central to Indigenous ceremony, festivals, and rituals that all four elders admitted to openly attending and enjoying (the dancing, singing, and praising elements of Indigenous ceremony) along with Native worshippers. This also supports Thiong'o (1986), Smith (1999), Dei (2000), and Battiste's (2002) discussion of orality and language as spiritual and central to the generational retention of Indigenous knowledges and identities for future generations. Additionally, Ms. Oriola and Mr. and Mrs. Awoniyi found it important that their children speak Yoruba fluently and insisted on only speaking Yoruba to their children in the household so that they would not lose their language. This is an example of how these community elders use their Indigenous language as a tool of resistance to colonialism and Eurocentric domination. As Ngugi wa Thiong'o (1986) reminds us:

> The choice of language and the use to which language is put is central to a people's definition of themselves in relation to their natural and social environment, indeed in relation to the entire universe. (4)

Additionally, Mi'kmaw scholar Marie Battiste and James Henderson describe the strong connection between language and spirituality as follows:

> Where Indigenous knowledge survives, it is transmitted primarily through symbolic and oral traditions. Indigenous languages are the means for communicating the full range of human experience and are critical to the survival of any Indigenous

> people. These languages provide direct and powerful ways of understanding Indigenous knowledge. They are the critical links between sacred knowledge and the skills required for survival. Since languages house the lessons and knowledge that constitute the cognitive-spiritual powers of groups of people in specific places, Indigenous peoples view their languages as forms of spiritual identity. Indigenous languages are thus sacred to Indigenous peoples. They provide the deep cognitive bonds that affect all aspects of Indigenous life. Through their shared language, Indigenous people create a shared belief in how the world works and what constitutes proper action … *Any attempt to change Indigenous language is an attempt to modify or destroy Indigenous knowledge and the people to whom this knowledge belongs*. (Battiste & Henderson, 2000: 48–50; emphasis added)

Indigenous language is the spiritual heartbeat or drum of a people and their culture. It connects and carries. It is the center through which the cosmological sphere is lived, passed down, and taught to subsequent generations because it holds cultural knowledge. Indigenous language is spiritual and can be read as a powerful form of collective self-affirmation that these community members engage in.

This group's insistence that spirituality was distinctly different from—although connected to—religion allowed me to pay closer attention to the complex entanglements and shared characteristics of community members' understandings of Indigenous Yoruba spirituality, versus the more Christian-based constructions of religion. The theme of a concealed Yoruba self (that is both spiritual and physical) that covertly operates inside and outside Christian demarcations of the spiritual, physical, and communal self recurred in my conversations with Yoruba elders and youth. This idea is significant because it allowed me to tease out the various meanings of the self that folks offered. Yet, I was cautious not to assume that a critique of church attendance or the revelation of concealed selves was somehow synonymous with an uncomplicated embrace of Indigenous Yoruba spirituality and practice. Constructions of self varied and were sometimes highly congruent with Indigenous Yoruba notions of being, yet at other times they were paradoxically and hegemonically entangled in colonial constructions of Indigenous Yoruba spirituality. One of the more overt constructions of spirituality overwhelmingly pivoted around many community members' assertion that involvement in Indigenous Yoruba spiritual belief and practice involved a lifestyle riddled with harm, evil, and moral decadence. I discuss this as an element of closeted life in the following section and refer to such constructions as "Discourses of Harm, Danger and Evil."

Spiritual Closets: Dichotomized Discourses of Harm, Danger, and Evil

The Christian versus Native construction of spiritual or religious identity was overwhelmingly present in my conversations with Yoruba community members. These constructions also heavily hinged on binarized character judgments, or notions of us versus them, where people who lived such a lifestyle were implicated in something that was assumed to be evil. All individuals were asked the same question regarding Yoruba spirituality: What are your thoughts or feelings about Indigenous or traditional Yoruba spirituality? Dichotomized constructions of Christian/Native and Us/Them resurfaced in many community members' responses to this question. For example, after, and despite alluding to the existence of a private spiritual self that exists outside her church life, Mrs. Awoniyi had the following to say about Indigenous Yoruba spirituality:

> Eh, ahh, I … ah, that to me … it's a no, no. I don't like it. I can give you an example, ok? Ah, at one time my Uncle, they wanted to give him a chieftaincy title. His wife called my mother [to tell her] that her brother is about to take a chieftaincy and my mother just ran there [to his house]. She said, "No, no, no!" The thing is, when you go in, you start eating with them. They have lots of different things, traditional things … if you go into the traditional things, it's not just having a chieftaincy title or being the Oba, there is more to it when you go inside. Whatever they are doing when you go there, that is what you will be doing too. To them it might be good, but to me … I don't like the way the Obas, the chiefs, the way they do their stuff. I come from a Christian family and we want it to stay this way. Because, at one point they wanted my father to become an Oba and he had to run away. He said he didn't like it and didn't want to be like them.

For Mrs. Awoniyi, the us versus them dichotomy is a truth that is real. In essence, it is a classic form of othering where, for her, "we" refers to the Christian dominant culture that holds normal values and beliefs, while "they" are the traditional(ists) who are different (read backwards), inferior, and not normal. "We" are Christian and therefore modern, progressive, and good; "they" are Obas (Yoruba word for King) and chiefs who do different and traditional things such as enticing you to eat with them. This form of othering places those who practice and believe in Indigenous Yoruba spirituality outside popular notions of normalcy, thereby stigmatizing them as abnormal. It is important to remember, however, that the type of othering that Mrs. Awoniyi engages in is one that is socially constructed, and bound up with imperialist constructions of the Indigenous Yoruba self as inferior, uncivilized, and backwards, notions that she has been socialized to think of as normal. Here, the reality of internalized oppression becomes a visible feature of Mrs. Awoniyi's self-understanding. This form of othering supports Pieterse's (1992) position that attributes of otherness can often be assigned—in this case it is the spiritual that evokes such categorizing—and doing so serves multiple functions for the labeling group, such as creating social distance; a claim to a privileged status; or it may also serve to assist in negotiating internal group relations by way of reference to the outsider group (230). The otherness of Indigenous Yoruba spirituality (and its practitioners) that pervades Mrs. Awoniyi's psyche serves all three of these functions. Her assertion of Christian identity, and emphasis on Indigenous spirituality as a "no-no," coupled with her insistence that it is a spiritual and social space that one should run away from clearly demonstrates her need to socially distance herself from Indigenous Yoruba spirituality. In addition, her repeated emphasis on "running away" from involvement in this spirituality suggests images of danger, peril, and consequential harm for her if she does not resist becoming one of them. This conceptualization operates to concretize and give legitimacy to her Christian identity as one that is good and safe when juxtaposed with the Indigenous one that is manufactured as dangerous, harmful and evil. Finally, Mrs. Awoniyi's reference to Indigenous spiritual believers as different and traditional fixes them in unevolved time and locates them as the outsider group, asserting her normalized position within Yoruba social society while claiming status within the dominant privileged Christian norm.

An enduring feature of community members' constructions of Indigenous Yoruba spirituality was what I term "discourses of harm." Within this discourse Yoruba elders and youth repeatedly offered anecdotal stories as examples of how people had been harmed by practitioners and believers of the Indigenous Yoruba faith. For them, these stories were the proof that substantiated how dangerous this Indigenous tradition was. When I asked Bisi to share her thoughts and feelings about those who practiced Indigenous Yoruba spirituality, she said that she was suspicious of them because "they were weird" and "wondered what they prayed to" because she was not sure it was God. In addition to her binarized construction of traditional practitioners as "Other," Bisi's doubt around who they pray to suggests that it is not God, which, for her, meant that they were odd and she was therefore justified in her suspicion of them. Such perceptions convey internalized colonial constructions of Yoruba Indigeneity as a pagan and therefore godless religion. The enduring Christian-based construction of Yoruba Indigenous spirituality as paganistic is set up to justify suspicion of the "Other." Since Christian constructions of God are monotheistic, any belief in Gods (or higher spiritual forces) that are not imagined according to Christian doctrine are set up as spiritually, morally, and socially inferior. The binaries of good/evil, safe/harm, heaven/hell, etc. that are fundamental to Christian theology, doctrine, and belief encourage a way of imagining spirituality as fixed and quite rigid. Here, Christianity proclaims itself as good, safe/heavenly, and of God. This then relegates the Other to a position of evil, harm/danger, and of the devil/hell its binarized opposite: Indigenous Yoruba (African) spirituality.

The suspicion of Indigenous Yoruba spirituality as harmful and dangerous was also voiced by Yinka, who distanced herself from this faith by adamantly denying the existence of any believers or practitioners in her immediate family, and instead insisting that her genealogy was exclusively Christian. In response to my question about traditional Yoruba spirituality, she had the following to say:

> Um … okay. If I think back to my Grandfather's generation, and even going further than that, they were Christians! Like not in the sense of being spiritual, or in the sense of going to herbalists and stuff. And when I say herbalists I mean like, um, voodoo doctors and stuff like that. No, he was a Christian. He didn't know how to read or write, so he didn't read the Bible. But he knew how to pray. And he knew that there was a God and he used to pray like that. Um … my sense of ah, people that go to spiritualists, or go to herbalists and stuff, those are people that … they're spiritual yeah. But they're not spiritual in a good sense. No I don't believe that they are. Some, some could … some … Ok, going to a spiritual doctor is perceived as being negative. Most times people that go to those kinds of things are people that want to do harm to other people.

The binaries introduced in Yinka's descriptions of Christian versus Yoruba spirituality conjure up and reproduce highly racist and Eurocentric constructions of "voodoo," the quintessential (demonized) representation of African spirituality, where notions of harm and danger are definitively undeniable in the European imagination. Yinka's remarks demonstrate that many Africans/Blacks are also affected by these constructions, images, and stories, through their continued circulation in mainstream discourses such as the media, for example. As colonized subjects, we may hold deeply internalized colonialist views about ourselves, our cultures, and our histories, albeit with differing consequences. These are the enduring traumas of colonialism. And this is evident in the stories of harm that both Yinka and Bisi relayed to me in our continued discussion of their views on Indigenous Yoruba spirituality. Yinka told me the story of a neighbor: an old woman who had attempted to kill her when she was very young in Nigeria and that it was her mother's church that had found the old woman out. In Yinka's words:

> I lived in Nigeria for three years from the age of two to five. So there were different apartments within the house but we had separate houses. And there, there was a lady that lived there, a nice lady, an older lady who um … my mom used to let us go play with her and she used to like me a lot. And my mom used to go to a church at that time, a Pentecostal church that was very spiritual. And there was times when they were doing like a revival at the church and the minister that day kept calling my mom's name, right. He was calling my mom's name like saying, "Yinka's mother" but my mom didn't take it in. But then somebody told her like, "aren't you the one they're talking about?" She's like ah, yeah. Then they're like, ok, why don't you go see this person. So she went to go see this person and when she did, right, [she starts to speak in a loud excited whisper] the person told my mom that somebody she lives near is trying to kill me! So my mom had to fast and pray. She had to fast and pray like I don't know for how long. But what had happened was, it was the lady's child that died. I used to go to her house, we used to live in the same place because everybody knows everybody in the same duplex. Yeah, so it's [traditional Yoruba spirituality is] usually associated with negative things. But I guess there's some people that could use it maybe for money, for monetary gain or something right. But it's not usually used for good things.

Similarly, the story Bisi told me involved danger, harm, and personal violation:

> My aunt, when we were in Nigeria, our house was robbed and apparently the people that robbed the house, they went to one of their oogun (Yoruba medicine) men and told him that they should make it so [that] whoever answers the door, [they] just let them in and take their stuff, and that's how they take the stuff. And that's exactly what happened. They came dressed up like they were visitors and cleared out the house. And then my aunt was so affected [by that] until she passed [away]. So when people say it's not real, yes it is because they wouldn't just keep talking about it if it wasn't, and especially because it happened with someone in my family. So that's what I think about Yoruba religion.

This is not to deny community members their agency by suggesting that these stories did not happen. In fact, for me, the question is not whether or not such stories are real. Rather, the issue is the manner in which these stories have become the quintessential representation of Indigenous African spiritual identity for Yinka and Bisi. For them, it seems these constructions of African spirituality have become the only valid understandings, which in turn get recirculated into the public imagination. It is not surprising, then, that Yinka and Bisi hold similar views because essentially, their perceptions are paradoxically theirs yet not theirs. In other words, because dominant discourses about Indigenous Yoruba spirituality are so hinged on the themes of harm and danger, that even folks who had more critical sophisticated understandings of Yoruba spirituality also felt the need to qualify their views with a harm disclaimer. For example, Sade Oriola argued that she knew many non-Christians who "worshipped idols" yet were still "good" and "very nice people who had good hearts." Similarly, Mrs. Fayemi took the position that it was not necessary to have to choose between Christianity and one's Indigenous spirituality, provided it (Indigenous spirituality) was used for doing good and not evil. In response to a question I asked about her thoughts on Wole Soyinka's spiritual beliefs, Mrs. Fayemi states:

> Yeah, his Orisa [a Yoruba deity] is Ogun. I think it's good! As long as you don't use it to harm anybody, just to protect yourself, I agree with Wole Soyinka. You can be a Christian too. As for me, I believe that we should hold onto our traditional culture. You know, we shouldn't say that because we are Westernized that we should just forget our traditional culture and rely only on Christianity or Islam. No. As far as you are not using it [Indigenous Yoruba spirituality] to do bad things, and you are just using it to protect yourself, to protect your family and you use it to do good to other people, there is no problem in practicing the two.

Disclaimers of harm such as Sade's and Mrs. Fayemi's are evidence of the profound extent to which discourses of danger and looming violence circulate in the public imagination about Yoruba spirituality. Mrs. Fayemi, in her awareness of this construction of her Indigenous spirituality, felt that voicing her support of it also meant proclaiming her disapproval of what it is popularly associated with: danger, harm, and violence. In this sense, Mrs. Fayemi is engaged in a psycho-spiritual act of resistance where, to her, Indigenous spirituality also plays a protective role beyond hegemonic notions of danger, harm, and evil.

These notions of danger, harm, and evil were ever-present when traditional Yoruba spirituality was brought up and discussed by other community members, suggesting that in their minds, these are characteristic of Indigenous Yoruba spirituality. For many folks, their salvation or guaranteed escape from such danger can only come in the form of Christian conversion and identification. Inversely, when themes of evil, harm, and danger arise in Christian doctrine or belief, they are assumed (read: constructed) to be coming from somewhere else outside Christianity, and are not deemed to be inherent to

Christian life, belief, or practice. In general, Yoruba folk held to the rigid binaries in Christian theology which dictate that values of righteousness, love, peace, and God's divine presence are the preserve of Christian religious tradition and no others, especially Indigenous traditions that continue to be set up as the classic antitheses to Christian monotheism.

Dichotomized constructions of Indigenous Yoruba spirituality as inherently dangerous, harmful, and riddled with evil generate oppressive yet powerful ideas that racialized Africans always do harm to one another conjure up images of the metanarrative of primitive tribalism. What is erased within this pathologizing discourse is the fact of colonial oppression, and how hegemonic constructions of Indigenous Yoruba spirituality are intrinsic to colonial ideology and European ethnocentrism. Eurocentric discourses of harm, danger, and evil are deeply embedded in historical amnesia and deny the fact that oppression is an integral part of colonizing and imperialist ideology and its related cultural, social, and legal institutions. These discourses of harm and danger have more contemporary manifestations in other social arenas such as the news media, where notions of danger, violence, and peril circulate about African/Black people. One of the most popular and enduring examples of this is the racist discourse of Black-on-Black crime in which media audiences are repeatedly fed images of gun-toting young Black men (read: gang members) that hurt, maim, and inevitably kill other young Black men of comparable rank. Similar to the discourses of harm, danger, and violence regarding African spirituality, discourses of Black-on-Black crime reinforce hegemonic notions of tribalism and innate in-fighting so that again, the impact of colonialism and imperialism are conveniently jettisoned, leaving the blame to be placed on the shoulders and souls of the oppressed: African peoples themselves. What is created and becomes normalized is a highly toxic anti-African climate that injures the individual and collective psyche. Within such a climate the open embrace of and identification with Indigenous African spiritualities is not welcome. And Indigenous peoples are thereby encouraged to deny, conceal, or completely distance themselves from Indigenous Yoruba (and other African Indigenous) faith because of its popular association with danger, harm, and evil. This discourse then becomes written onto both the individual and collective racialized African body. Meanwhile, the sources of such dominating constructs remain invisible and do not enter the public imagination or psyche with even a fraction of the disdain that Indigenous African spiritualities do.

Not only does the circulation and hegemony of these discourses pathologize African/Black peoples and their Indigenous spiritualities, they also serve the function of eclipsing the significant stories of survival, healing, self-determination, self-love, empowerment, and resistance to colonial and imperial oppression that are embedded in Indigenous Yoruba (African) knowledges and practices. The invisibility of these stories is another layer of the individual and collective psychic and spiritual wounds that Indigenous people suffer when they have deeply (generation after generation) internalized such pathologically racist notions of themselves, their culture, and their collective Indigenous identity.

Discourses of Evil and Idol Worshipping: Demonizing the Yoruba Orisa

> The devil is not the terror that he is in European folk-lore. He is a powerful trickster who often competes successfully with God. There is a strong suspicion that the devil is an extension of the story-makers while God is the supposedly impregnable white masters, who are nevertheless defeated by the Negroes.
>
> —Zora Neale Hurston, *Mules and Men* (1978): 230

I begin this section with a quotation from Zora Neale Hurston because her analysis of the devil quieted some of the unsettled tensions and questions I had been struggling with around Christianity and Indigenous Yoruba spirituality. With time, I noticed that this struggle often pivoted around one particular figure that was repeatedly evoked as evil incarnate: the devil. In many Yoruba Christian communities this figure is called Esu. At various churches and especially in fervent prayer, Esu's name is often invoked in concert with what people prayed would not happen to them and their loved ones, that is, sickness, failure, loss of employment, death, poverty, etc. Esu is also largely blamed for any form of evil or negativity in people's lives. In the many Yoruba churches in the city I have attended, Esu always seems to figure in the same way: as an ominously lurking force that as a committed Christian, one must to do everything possible to rebuke and stay away from. As Yoruba Christians, goodness, benevolence, and spreading the Gospel are your personal and social charge, and are often emphasized by a preoccupation with the need to conquer and destroy the existence of malevolent and demonic forces. Such forces are always evoked in connection with Esu's title and name. At home during family prayers, Esu is rebuked and cursed. In church and at other social functions, Esu is rebuked and cursed. The message is clear: Esu is the consummate emblem for sin and evil and every effort should be made to stay away from "him" and all that "he" represents.

In this project all community members repeatedly referred to Esu and other Orisa/Yoruba deities as "idols" and any form of reverence or embracement for them was viewed as "idol worshipping." This in turn translated into charges of not knowing God. Pejorative words such as "pagan" were used to describe believers and practitioners of Indigenous Yoruba spirituality. When I asked Mrs. Olusanmi (Mama Niyi) to share her thoughts on Indigenous Yoruba spirituality, she said it was idol worshipping and went on to explain that, as a Christian, she was brought up to not like it and not deal with the people who practiced it, saying:

> Christians don't deal with those people. Even when they are dancing we Christians will pretend not to see them anyways because the way I was brought up … I was brought up I won't say to hate them but I was brought up not to like them … The thing I grow up with is still with me. Even up to now, like here, I look at Halloween as something that is similar to the traditional beliefs back home. I don't like it. I don't participate in it.

Similarly, Mr. Oladiran named Esu as "the Yoruba idol that is the devil." When I questioned his assertion that Esu was the devil, and asked him how he came to know this, Mr. Oladiran responded:

> Okay, in the Yoruba culture, Esu happens to be representative of the devil because even in the idol worshipping of Esu far back home in Africa, you will find out that it is always placed in the centre of the … wherever they [Native worshippers] want to do their festival. Even within the complex of the dominion where they are worshipping in, it's never put inside, it's always outside because the idea is that you don't want the devil in your home because he can only do harm to you. So whatever they the native worshippers want to do, they do it outside. So in the context of us Yorubas, Esu is the devil.

Indeed, within this discourse it seems that all Yoruba deities are, in fact, idols. There is a complete disregard of the historical fact of colonization and how it has influenced Indigenous self-concepts to denigrate Yoruba spirituality. Colonialism deliberately positioned Yoruba Indigenous knowledges as backwards, uncivilized, idolatrous, and therefore inferior. In fact, colonialism has constructed many Indigenous knowledges as non-knowledge and therefore intellectually, socially, and morally inferior. Meanwhile, Euro-Christian ideology is situated as the divinely sanctioned norm, emphasizing the first and second commandments (as delivered by Moses) which stipulate that making any image of gods or worshipping any other God than the Christian one is forbidden. Hence, the culturally specific Euro-Christian construct of idol becomes a universal norm by which Other spiritual traditions are measured and judged. These traditions—and Indigenous ones in particular—can never quite measure up because Christian philosophy is profoundly invested in its own superiority. In Euro-dominant cultures, such constructs have been widely naturalized as universal spiritual doctrine. So, if Christianity asserts itself as constructed as monotheistic, Indigenous traditions are pathologized as polytheistic and therefore paganistic. In a Christian model, the existence of more than one god in a spiritual tradition invokes accusations of idolatry. Other Euro-Christian binaries such as God/devil, good/bad, progress/backwardness, civilization/primitiveness, holy/demonic, and, ultimately, Christian/pagan emerge as natural and normal within this model. Within such a frame, there is no tolerance for Other approaches to spirituality, and Yoruba Orisa (deities) such as Esu are rewritten within the naturalized frame of Eurocentric discourse as idols. As the only Orisa that appears in the Yoruba Bible, Esu becomes the unfortunate and most popular casualty of Euro-colonial Christianity.

The Politics of Exile: Spiritual Closets, Secrecy, and Injury of the Soul

Discourses of harm, evil, and idol worshipping that dichotomize and demonize Indigenous Yoruba spirituality are, in fact, manufactured colonial agendas where erasures, amputations, and silences of Yoruba identity reign.[22] The Yoruba figure of Esu is misinterpreted as an idol by most Yoruba elders and community members and was ultimately believed to be the devil by them. It is within a Yoruba Christian community that Esu has come to be understood as the devil. However, the reality is that in this climate such a need is often inextricably connected to dominant notions of respectability; therefore, being a respected member of the Christian Yoruba community has meant aligning oneself with and/or internalizing dominant Euro-Christian constructions of normalcy. Jacqui Alexander (2005) reminds us that the notion of respectability is hegemonic and reflective of conventional (colonial/imperial) ideologies, values, and beliefs, including and especially religion. To be respectable is to be a member of the dominant culture and to therefore claim a dominant religious identity. In this project that centers diasporic Yoruba, that would be Christianity. However, such outward religious allegiances do not completely erase one's Indigenous consciousness. Rather, they push Indigenous ways of being and knowing to the peripheries of Yoruba life in the diaspora to a space where they are concealed and engaged in secret. In this way, both Yoruba

Indigenous knowledges and their practitioners are forced into spiritual closets, while cleaving onto conventional outward claims of Christian identity and religious normalcy.

For instance, Dele discusses the social and spiritual significance of the head in Yoruba culture, and how this is particularly emphasized in an Indigenous Yoruba ritual that was performed on his own head. While he states categorically that he and his family are Christians, in the same breath, he also recognizes that connections to believing in and practicing Indigenous Yoruba ritual endure. Dele explains:

> When I was coming to Canada, my Grandma, she did something to my head. She called some man that did something on my head as if [it was] some kind of protection. That was when I was coming here [to Canada] like when I was 12 years old. I remember my head being shaved and someone making some kind of … I'm not sure what it was on my head but I didn't dare say anything about it either because my Grandmother was the boss lady [laughing]. She was the woman of the family kinda thing. And when I got here [to Canada] I remember some other older lady asked me and my mom what happened to my head and my mom was struggling to explain and the lady was like "Oh, okay I get it." [laughing]. I figured she was another older Yoruba lady that probably knows about it. I actually didn't know what was … ahh … going on. But they just did with my head what they did … what they wanted to [laughing]. But it's also … it's embarrassing because we're Christian, you know what I mean? Like our family is known to be Christian and yet maybe there still remains, especially with older people like our Grandparents, they still have their little attachments to their um … what's it called? The Orisas and stuff.

First, African/Black women's work of protecting their children and families is a labor of love, a spiritual labor that Patricia Hill Collins (1990) identifies as a core theme in Black feminist thought. And this protective spiritual labor is what Dele's mother and grandmother were performing for Dele in the ritual that involved his head. Second, Dele's experience highlights the enduring survival of Indigenous Yoruba spirituality, particularly among the older generation. However, his construction is also situated within a framework of shame, which unsurprisingly encourages an outward claim to a Christian identity. Dele's brief but poignant allusion to Indigenous Yoruba spirituality is worthy of attention and, I would argue, reflective of its hidden, subsumed existence with/in the larger milieu of Christian life among many diasporic Yoruba youth. In other words, Dele finds the ritual performed on his head to be shameful because he understands that it is not a Christian ritual and, despite his astute recognition that such an event was carried out to protect him, it does not take away from his embarrassment and shame around it. Such notions of shame are endemic to spiritually closeted Indigenous life and key contributing factors that force Indigenous Yoruba spirituality and its practitioners into a deeper spiritually closeted existence.

My argument of closeted African spirituality builds on Zora Neale Hurston's (1978) classic ethnographic study of African American folklore, where she explores how African informed spirituality is cut in the woes of secrecy and therefore forced to operate in secret. She argues that this is so primarily because it is not the religion of the nation:

> Nobody knows for sure how many thousands in America are warmed by the fire of hoodoo, because the worship is bound in secrecy. It is not the accepted theology of the Nation and so believers conceal their faith. Brother from sister, husband from wife. Nobody can say where it begins or ends. Mouths don't empty themselves unless the ears are sympathetic and knowing. (195)

Hurston's discussion shows us that a closeted practice of African Indigenous spirituality reigns above even some of the most intimate relationships in one's family and community. The need to keep one's spiritual practices and beliefs secret, or in the closet, is necessary to avoid ostracization and accusations of practicing witchcraft, juju, or voodoo, all of which emerge from Eurocentric discourses where once again, harm, danger, and evil are assumed to be the norm among Africans/Blacks. This discourse of Indigeneity as a pathological, diseased, and abnormal spirituality and fact of Black/African life makes it socially (im)possible for many to disclose belief of and involvement in. For many colonized individuals to claim a Yoruba Indigenous spirituality would be to suggest that one is actually pathological and involved in practices that are harmful, dangerous, and, essentially, not of God. The fear is that these (normalized Euro-Christian) beliefs are ones that one's family and closest friends will also hold to be true and maybe, therefore, risk losing the love, honor, and respect of those who matter the most. Essentially, to openly embrace one's spiritual Indigeneity means that you are running the risk of losing your respectability and committing social suicide.

Some may reject or not be convinced by my argument of spiritual closets with the opinion that spirituality is an individualized affair, explaining why community members did not want to speak about it; or that they may have a particular unspoken interpretation that is not necessarily akin to a closeted knowledge. While it has been argued that not wanting to speak openly about one's individualized spirituality can in itself be part of a person's spiritual belief, it is crucial to remember that I make the argument of spiritual closets within the context of colonial and imperial oppression, not forgetting that these forces are pervasive and that people navigate their daily lives in negotiation with these realities. Linda Smith (1999) cautions against the temptations of adopting "misty-eyed" models of Indigenous spirituality that are not grounded in the everyday material experiences of Indigenous people, namely, the oppressive hegemony (spiritual and otherwise) of colonialism and imperialism (12). Community and collectivity are core elements of Indigenous spiritual philosophy, belief, and everyday lived practice. From an Indigenous perspective, it is unheard of to not want to openly practice, speak of, or engage in what is for Indigenous peoples the source of their individual and collective being, unless there are strategic and/or protective reasons for such secrecy. These reasons include the Eurocentric and pathologically colonialist constructions of Yoruba (African) spirituality that deeply pervade the moral, social, and institutional fabric of Western society. It is no secret that African Indigenous spiritualities are often relegated to the lower rungs of colonial stratification, thereby devaluing Indigenous ways of knowing and being. Openly embracing this spirituality carries little if any currency in terms of social mobility, institutional respectability, and overall privilege within dominant Euro-Christian colonial culture. Additionally, to locate Indigenous spiritual practice and belief solely at the level of the individual is to adopt New Age interpretations of spirituality that are not grounded in the daily material realities of Indigenous peoples and their worldsense. What is forgotten is the interdependent and multi-layered nature of Indigenous spiritual practice/knowledge, where the individual is but one inextricably linked layer of a larger cosmological reality. Sobonfu Some (2003), Indigenous spiritual teacher from Burkina Faso, elaborates:

> Day by day we work to maintain our state of grace. We do so not only as individuals, but also as a part of several interconnected circles of support. When we fail, the work of coming back into grace is something we cannot accomplish by ourselves; it requires the participation of others.
>
> The cosmos, the universe, is the largest circle to which we belong. This is the realm of Spirit, of goddesses and gods, of our ancestors. The next circle comprises the planet we live on, Earth. This is the place of air, water, fire, soil, stones and trees. Then comes our country and culture. Nearer to us is the circle of community, the friends and coworkers and others with whom we share our daily life. Our extended family makes up the next smaller circle, including our parents, children, brothers, sisters, uncles and so forth. Lastly, I think of the circle of intimacy, which we share with a spouse or partner. (23–24)

From within the context of colonial and imperial oppression, New Age constructs of spirit appropriate the realities and experiences of Indigenous peoples, displace them outside of the Indigenous communal framework to then romanticize and reframe them in a depoliticized neoliberal context that privileges White individualism. In short, the argument that Indigenous spirituality be seen or approached as an individualized affair does not explain Yoruba elders' and community members' responses in this text, not only because collectivity is a central feature of Indigenous spiritual life but also because this is a project that infuses the spiritual with the political. In so doing, social relations of power are highlighted particularly with respect to how colonialism, slavery, and oppression have hegemonically (re)(con)figured Indigenous spiritualties to operate in secret, or are corrupted in a neoliberal framework.

I now return to the Orisa Esu who is an important figure for this text. I also return to Yoruba worldsense to situate this deity within an Indigenous and anticolonial context. As mentioned earlier, Yoruba peoples approach human life as part of the eternal existence of spirit that resides in *Orun* (the Otherworld), our everlasting home. This is illustrated by the popular Yoruba proverb, *Aiye l'oja, orun n'ile* (this world is a marketplace, the Otherworld is home). In essence, this means that human life is understood to be a journey to the marketplace we call Earth where both matter and spirit are infused to create the various forms of life that exist here on Earth. In Yoruba cosmology, the Orisa Esu has a vital role because of this deity's two primary responsibilities: messenger and owner of the crossroads, and the keeper of Ase. As owner and messenger of the crossroads, Esu ensures that links between the spiritual and physical members of the African community remain intact. That is, that one's community of ancestors, human beings, and the unborn remain linked and connected. In doing so, Esu acts as a necessary spiritual medium and messenger. However, it is Esu's second charge that I focus on here.[23]

As keeper of *Ase*, this Orisa is the keeper of life-force energy and power (Abiodun, 1994), what I refer to as "the essential breath of life." The significance of such a charge cannot be underestimated because it means that this Orisa plays a central role in power and spirit and all matters pertaining to this: *everything*. It is difficult and, frankly, quite overwhelming to attempt muddling through the kinds of consequences such a colonial transposition means. However, I can say that demonizing Esu effectively means demonizing African Indigenous spirituality and in that sense Esu becomes both the literal and symbolic representation of how African spirituality has been reconstructed, reframed, and misunderstood. Hence, the internalized words and ideologies that many of the Yoruba elders and youth use in relation to Esu echo the Euro-colonial Christian ideologies and discourses repeatedly evoked when discussions about Indigenous African spirituality arise. Notions of harm, danger, evil, idol worshipping, and pagans are used interchangeably by Yoruba community members and elders to denote both Esu and Indigenous African spirituality as demonic. To be coerced to engage one's Indigenous spirituality in this manner, that is, to internalize, deny, engage in secret, or completely disengage with this aspect of one's culture is to deny, hide, and disengage with aspects of one's Indigenous African self. It is to stifle what is quite literally a powerful energy or force of life that has the potential to be one of the most spiritually nurturing and liberating parts of ourselves. This is especially the case in colonially oppressive climates such as that of North America. In turn, this has enduring generational consequences that are both spiritual and material. In essence, the metaphoric and literal transposition of Esu and Indigenous Yoruba (African) spirituality is an acute form of soul injury, or, what Black feminist scholar Patricia Williams (1997) refers to as "spirit murder":

> I see spirit-murder as no less than the equivalent of body murder. One of the reasons that I fear what I call spirit-murder, or disregard for others whose lives qualitatively depend on our regard, is that its product is a system of formalized distortions of thought. It produces social structures centered around fear and hate; it provides a tumorous outlet for feelings elsewhere unexpressed ... We need to see it [spirit-murder] as a cultural cancer; we need to open our eyes to the spiritual genocide it is wreaking on blacks, whites, and the abandoned and abused of all races. We need to eradicate its numbing pathology before it wipes out what precious little humanity we have left. (235)

Injury of the soul occurs on both the individual and collective levels, inflicting wounds to the psyche that induce a type of *spiritual* psychic violence.[24]

The pragmatic necessity of closeting one's Indigeneity and spiritual beliefs is burdensome and amputative to one's soul. Such wounds are injurious to the spirit in the sense that closeted forms of spirituality entail a denial or eclipsing of the self and create internal fragmentation. These denials of the self can be traced back to the Eurocentric insistence on singular notions of normalcy. The deleterious effects of this colonial fixedness do not allow for the body, mind, and soul to coexist in multiple and diverse ways. Again, this has enduring consequences that are both spiritual and material for subsequent generations. Unfortunately, for these younger generations, access to Indigenously anchored constructions of Yoruba cultural practices are generally stifled and blocked. This allows for the more popular Eurocentric constructions of Indigenous spirituality to propagate both within and beyond the Yoruba psyche and imagination. This is where the real spiritual harm exists. Contrary to what popular culture would have us believe, the real danger to one's person, soul, and community lies in the singular and hegemonically fixed notions of spiritual normativity that are dictated by Euro-Christian belief and discourse.

Notes

1. Sade Oriola, interview, 10 March 2007.
2. Mrs. Olusanmi, interview, 22 February 2007.
3. Mrs. Awoniyi, interview, 24 February 2007.
4. Mr. Awoniyi, interview, 24 February 2007.
5. Ibid.
6. Mr. Fayemi, interview, 25 March 2007.
7. Mrs. Fayemi, interview, 25 March 2007.
8. Mr. Oladiran, interview, 4 March 2007.
9. For extensive critical discussion and interrogation of respect for elders and seniority in the Yoruba community, see Chapter 5.
10. Mrs. Oladiran, interview, 4 March 2007.
11. Ibid.
12. Ibid.
13. Dele Oriola, interview, 10 March 2007.
14. Niyi Olusanmi, interview, 22 February 2007.
15. Ibid.
16. Bisi Awoniyi, interview, 24 February 2007.
17. Seun Fayemi, interview, 25 March 2007.
18. Ibid.
19. Yinka Oladiran, interview, 3 March 2007.
20. Tunmi Oladiran, interview, 4 March 2007.
21. Tunmi's aspiration to continue working at the Jane and Finch community is admirable given that this community is one of the priority neighborhoods in Toronto, Ontario. With a focus on community development and support it is comprised of many racialized immigrants and first- and second-generation families living in systemic poverty. This community is often stereotyped in the media as one of Toronto's most dangerous neighborhoods, especially given that a substantial number of residents are racialized, poor, and working class, particularly African Canadians.
22. See Chapter 3 for a more comprehensive examination of how the Orisa Esu was mistranslated as an "idol" and erroneously recast "the devil."
23. See Chapter 3 for a more detailed discussion and analysis of the Orisa Esu's Indigenous role as keeper of the crossroads and messenger.
24. I build on and include a spiritual dimension to Patricia Williams's (1997) discussion of "psychic violence" (233).

· 4 ·

AT A CROSSROADS

Esu, Language, and the Politics of Critical Spiritual Literacy

> I have found in my work, for example, that where a narrow kind of Christianity has been instilled, people accept that they have been born evil. This view infiltrates the way people look at each other. "We are basically evil." The battle against our nature never ends. This belief automatically limits a person's abilities to come back into grace. It's as if one's wings have been clipped before she can fly. It takes people out of the state of grace in which all babies naturally arrive.
>
> —Sobonfu Some, 2003: 27–28

The discussion here rests on how the sacred is embedded within and written into our everyday experiences and lives. More specifically, it spotlights how Yoruba elders and community members engage with and read the sacred in their daily lives. As mentioned earlier, this notion of literacy was inspired by Frederic and Mary Ann Brussat (1996) who define spiritual literacy as "the ability to read the signs written in the texts of our own experiences ... and find sacred meaning in all aspects of life" (15). Such a take on literacy extends beyond conventional Eurocentric notions of being able to read, which are overwhelmingly confined to orthography and comprehension of letters on a page. However, without a critical component, spiritual literacy cannot adequately address inequities that arise from oppression. For example, as discussed earlier, the Brussats identify worldview as the most significant block to spiritual literacy, and they specifically emphasize that seeing the world as "Devil-ridden, doomed and dangerous" (33) is to embrace a worldview that blocks one's ability to engage, read, and commune with the sacred in daily life. Yet such a perspective is strongly similar to Christian worldview and how some Yoruba elders and community members constructed their Indigenous spirituality in this manner. If left without a critical component that addresses inequity, the question as to how and why Yoruba peoples construct their Indigenous spirituality as dangerous, evil, and devil-ridden, while Christianity is posited as good, heavenly, and of God remains unattended to. If left this way, this leaves one vulnerable to uncritical speculation, which is likely to be shaped by racist stereotypes and hegemonic hierarchies. For this reason, the significance of *critical spiritual literacy* as a conceptual tool is highlighted, to help us better understand the value of Indigenous worldsense/cosmology as sacred and the dangerous implications of assuming that all worldviews can be measured, judged, and effectively interpreted according to Eurocentric worldview and values. Moreover, as a prime communicative vehicle through which knowledge and relationships are built and sustained, language is central to Indigenous worldsense and the ability to read the sacred in one's day-to-day life. In this manner, language is also central to critical spiritual literacy. To illustrate, I use the example of the iconic Black feminist film, *Daughters of the Dust*. In the critically acclaimed independent movie, filmmaker Julie Dash highlights the importance of language for African/Black people, noting that remaining true to the rich language of the Gullah people was priority for her in the film. Many Blacks who watched the film felt that the Gullah language was not a problem and therefore resisted the suggestion of incorporating subtitles (Bobo, 1995). However, there was expressed concern that subtitles were needed so that Whites would be able to understand the film. Julie Dash's position on the issue was in keeping with many Blacks who believed that there was a certain cultural dignity and discernment maintained in the insistence that subtitles not be included, and that viewers should patiently sit with the film. For Dash, it was important to respect and remain true to the linguistic integrity of African-diasporic culture and the overall story of *Daughters of the Dust*:

> To tell the truth, I had problems with *Miller's Crossing*. It made me realize that I've done that all my life, pushed through on accents until I understood them. Why is it with *Daughters of the Dust* that people seem almost offended by it? When they bring it up, I tell them, "Release on it, you'll understand it in a minute." You may not understand every sentence but you'll surely get the general idea, the sensibility of the whole thing. (Bobo, 1995: 188)

The significance here is that Dash's position demonstrates how language and literacy shape ideas about one's identity. In this particular example, Dash insists that the audience read the on-screen material on its own terms instead of being fed and/or distracted by standard English subtitles. Also implicit in this debate is the assumption that the language and culture of the Gullah people can be easily translated into standard English. This is not to say that translatability is impossible or undesirable. Rather, the challenge here is the assumption that standard English can effectively and equitably express the ideas, norms, idioms, and symbols of another culture without significant meaning from said culture being lost, particularly where Indigenous societies are concerned. This problem is what Indigenous scholar Marie Battiste (2000) refers to as "the Eurocentric illusion of benign translatability," where European-centered cultures do not give serious consideration to the differences between worldviews and languages and instead assume that they can be easily understood, and therefore translated without misrepresentation or impairment (79–80). Hence, critical spiritual literacy places as central the historical and contemporary power imbalances that have created and sustained a hierarchy in terms of how various worldviews are ordered, constructed, and valued. In this way, the concept of difference and the uniqueness of various worldviews is significant. Indicators of oppression, inequity, and the legacies of imperialism and colonialism are recognized as the key forces responsible for the hegemonic and hierarchical ordering of worldviews, and they are approached as foundational in structuring blockages to one's ability to acquire a spiritual literacy that is critical. Accordingly, critical spiritual literacy is used to fill in crucial gaps that have remained unaddressed, as well as to address the following question: What are the challenges of learning Yoruba Indigenous knowledges in a context that pathologizes African Indigenous culture and spirituality?

In essence, critical spiritual literacy centers the use of Yoruba Indigenous knowledges as sacred, and also considers the historical and contemporary power inequities that challenge the learning and use of Yoruba Indigenous knowledges by elders and youth in the diasporic Yoruba community. These inequities are colonial and I therefore place extreme importance on Indigenous people's abilities in terms of their freedom to commune with and recognize the sacred in daily life. Using a critical spiritual lens involves a deep appreciation for the connections between language, cosmology, and one's individual and communal identity. It ultimately requires a deeper exploration of the assumptions embedded in dominant and Indigenous cosmologies, while addressing the consequences of taking Euro-Christian discourses and ways of reading the sacred as universal. With this lens, I discuss how notions of cosmological imperialism and cosmological difference arose from interviews with diasporic Yoruba community members and Black/African feminist theory. These terms are subsets, or smaller components of the larger idea of critical spiritual literacy, and I discuss these smaller components by highlighting the Orisa Esu as my entry point. I do so because Esu is an Indigenous figure that was repeatedly referenced in Yoruba elders' and youths' accounts of religion and spirituality. Of note is that the figure of Esu is literally and symbolically central to Yoruba Indigenous culture and cosmology.

Demonizing Esu and the Problem of Amputative Ambivalence

> Well evil … evil can be of different categories right? Okay, the belief is that Satan is evil in the Christian realm, that [he was] one of the disciples of God that misbehaved and is trying to tempt people in the world. And then, in our traditional way they call Esu eh, evil too and he's a deity too; he's a god. It's um, it's like an intermediary between God and people. So he can … you know, he's a trickster in the Western world [laughing]. They see the red ties on one side, they see the black or something like that on the other. So basically there are two different types: Christianity and the traditional. (Sade, Yoruba elder)

Sade discusses two figures in the Euro-Christian and Indigenous Yoruba realm: Satan and Esu. At first glance, one may assume that she is suggesting these two figures are equivalent. However, this is not the case. Rather, I believe what Sade means is that she is pointing out the way Esu is (mis)understood by some (read: constructed) as the equivalent of Satan. Yet this deity is in fact more than that if one considers the roles and responsibilities of this Orisa within Yoruba worldsense. For instance, Sade notes that Esu is a trickster deity and points out the complicated dualistic character of this Orisa. It is red on one side and black on the other—where his or her[1] dualism is not binarized, but complementary (i.e., male and female, red and black, hot and cold, etc.) and intersecting.[2] In essence, Sade is identifying some of the differences between the Indigenous Yoruba and Euro-Christian metaphysical systems and suggesting that these differences are important. If this particular Orisa is demonized among the majority of the Yoruba elders and youth in the larger context of Yoruba Christian social life, this is due in large part to the generational internalization of Eurocentric Christian discourses. As previously mentioned, such discourses have hierarchically ordered Yoruba Indigenous knowledges as backwards, inferior, and uncivilized. However, interestingly, while carrying out extensive research and exploring critical scholarship on Indigenous Yoruba spirituality, these knowledges and Esu in particular were discussed by critical Black feminist and African scholars in quite a different way (Abiodun, 1994; Abimbola, 1977; Bakare-Yusuf & Weate, 2005; Oyewumi, 1997). In this scholarship, Esu was an Orisa, a cunning and central Yoruba deity that was the owner of the crossroads and keeper of *Ase*, the power and essential breath of life. I found myself in somewhat of a quandary because I was not able to reconcile these polarized constructions of Esu. While the research I completed showed that undoubtedly this was emblematic of being a colonized people where many had converted to Christianity, it did not speak to the present-day tensions I was feeling around how Esu had been positioned and was still demonized in contemporary Yoruba Christian life—both socially and spiritually. I was stumped, at a crossroads, ironically enough. And traversing back and forth between Esu, "the devil," and Esu, the central Orisa and keeper of Ase was a "nervous condition" that did not feel healthy (Dangarembga, 1988; Fanon, 1963). However, while being at a crossroads is an important reoccurring juncture in life, it is a site that one visits; it is only temporary. It is temporary because the crossroads propels you into making a choice. But yet again, the irony of this ambivalence is that paradox, contradiction, elusiveness, freewill to choose, and ambiguity are also domains of Esu the Orisa, the deity of the crossroads that sits, stands, and traverses between and with/in the material and spiritual worlds.

What this narrative highlights is the problem of congestion, and that blockages to one's Indigenous worldsense can manifest as a type of ambivalence, or a tension that operates as a metaphor for the larger problematic of religion as a site of contestation. Therefore, my aim here is to explore and tease out Yoruba community members' understandings of their Indigenous knowledges through the figure of Esu. As a profoundly demonized Orisa, Esu is unique because this deity is symbolic of how African Indigenous spirituality and Africans/Blacks have also been demonized and pathologized by imperial and colonial hegemony. Again, Esu is an important deity not only because of this Orisa's key role and function within Yoruba cosmology but also because Esu was the deity spoken of the most by diasporic Yoruba elders and community members. In this sense, Esu is the deity that brings the challenges of learning and engaging Yoruba Indigenous knowledges in Eurocentric contexts to light. Upon deep reflection as to why the Orisa Esu was repeatedly mentioned by Yoruba folk, it became clear that it was not coincidental because Esu is the only deity who was transposed into Christianity via the Yoruba Bible and Yoruba social life as the devil and archetype of evil. Given this, I situate the tension between the Yoruba worldsense and Euro-Christian worldview through two key figures: Esu and Satan.

Contemporary implications of cosmological encounter (such as blockages to spiritual literacy that are informed by the amputation of one's African Indigenous knowledges) have been paid scant attention to date, scholarly or otherwise. Yoruba elders' and youths' ambivalence around the figure of Esu is symptomatic of Christianization (proselytization and conversion) that continues to be a larger dimension of religious fundamentalism and globalization. Again, the existing consequences of this reality among the Yoruba diaspora in Canada is not paid much attention. Moreover, while on the one hand the overwhelming number of people converting to Christianity can be explained through a Marxist conceptualization of religion that posits that it is latched onto in times of social rupture (i.e., globalization), here I focus on how religion and spirituality become sites of contestation, as reflected in many community members' understandings of the Orisa Esu. I argue that such constructions are both informed by and operate as cosmological imperialism. Specifically, I explore how in this context this type of ambivalence manifests to block spiritual literacy and further the continual dominance of Eurocentric constructions of Indigenous worldsense. Sade gives a powerful account of Yoruba worldsense as important knowledge that is necessary to retain and pass on to the following generations:

> For me, yes I am in Canada, and what we call, *ilu oyinbo* [white people's country], but as Africans, we see the world in a different way. Um, we have ah, those that we can see, you know like us humans or people. But we also have those we can't see, at least not with our eyes. And it is all here too, it is all here; a part of the world no matter where you come from or go. This is what we Yoruba believe—and it is all owned and created by God, the highest power. This is very common knowledge for us Yoruba, and most Africans I know anyway. My sons, they know this too, even though we are here they know this. Like, no matter what, for me, they [our children] should know about the culture. It's very important because if you don't—like if they don't know about their culture, they will get lost. And everybody have their own culture. No matter what, even if you integrate into the Western world you still have to know your background, you still have to know your culture, and you still have to know that the way we do things, if it's different it's okay if we know God in our own way. So, it's important for me, but I don't know about others because some of our kids here don't know. But then, I know that definitely if you let them know about it, like if you teach them then they will learn and they will know.

Similar to Sade, Mr. Fayemi made distinctions between Western and Yoruba religious perspectives of the world and staunchly defended Yoruba worldsense as valid, too:

> I believe in my own culture and traditional values which is Ifa. In this Ifa, we Yorubas worship God through this and at the same time we worship Osun and then we worship the God of iron which is Ogun, yeah. You see, all of these small gods, they are parts of nature. You know, Ogun is the forest and Osun and Yemoja are water. And Sango is rain and thunder, and I believe in that! These traditional ways teach us to respect nature because God created it all! If you buy a new car, you [are] supposed to slaughter a hen in order to prevent evil occurrence to you, you know. Which is what I…it's a matter of spiritual belief! Then you believe in something, that if I do this, this will not happen and it's not going to happen! There are all types of powers around us, and us Yoruba we know that these things work! Your mind is your religion. But religion here [Canada] is not the same. We have been told that these ways of doing things is not civilized. That's why I told you I don't go to church. But my wife and children go and I don't mind. You know, it helps them, to see the world the way we do here. We are here and it helps them.

Here, Mr. Fayemi's analysis of Western worldview indicates his understanding that it is important to know this view of the world as a strategy for survival to navigate living in it. Mr. Fayemi felt that his wife and children did this by going to church. In this sense such a strategy can also be read as a way of teaching their children the strategy of surviving in this land. However, when I extended our discussion of the Yoruba Orisa to Esu in particular, Mr. Fayemi did not doubt that Esu was the devil:

> No. Esu is not an Orisa. Esu is the devil. That is the Yoruba word for devil or Satan. Esu is a devil. When they say somebody is Esu, it means it's a bad person, he's working for devil. They use it in so many ways. They can use it to describe some human being … Yes that particular word is for evil things, and devilish things. They can use it to describe anybody; they can say, "this girl is Esu!" [laughing]; that if you are causing problems that's what it means. It's not Orisa, whosoever is causing—even when a goat or dog is eating your maize you say, "this goat is Esu!"

Drawing on Yoruba community members' conceptions of Esu, I present three components that are central to critical spiritual literacy that focus on (1) cosmological imperialism, (2) cosmological difference, and (3) how the Yoruba and Euro-Christian cosmologies encounter one another. First, I employ a discussion of cosmological difference by outlining Esu's role and function in Yoruba cosmology and philosophy. Second, I engage a historical analysis of how blockages to critical spiritual literacy emerge in an African/Black feminist analysis of Samuel Ajayi Crowther, the key figure who translated the Bible into Yoruba and Esu unto the devil. This component is particularly significant because it maps how hegemonic constructions of race and gender recur throughout Crowther's translation. I build on this by showing how these constructions produce dichotomized and essentialized versions of Esu that then fix this deity as hegemonically masculine and quintessentially evil. Last, I draw some conclusions as to what the implications of this congestion are and why it is a problem for diasporic Yoruba as individuals and communities to bear such a burden.

Cosmological Difference: A Struggle Between Two Metaphysical Systems

> The term "worldview," which is used in the West to sum up the cultural logic of a society, captures the West's privileging of the visual. It is Eurocentric to use it to describe cultures that may privilege other senses. The term, "worldsense" is a more inclusive way of describing the conception of the world by different cultural groups ... [and] will be used when describing the Yoruba or other cultures that may privilege senses other than the visual or even a combination of senses. (Oyewumi, 1997: 3)

I situate my conception of cosmological difference under the larger rubric of cosmological imperialism and draw on Audre Lorde's (1994) discussion of difference. Lorde argues that it is imperative to move beyond mere tolerance or "pathetic pretense" that our differences do not exist. Rather, she encourages us to view difference as a primary site where we acknowledge our strengths as different yet equal. In this way we can garner the power to seek new ways of being in the world (1984: 111). The philosophical differences between Yoruba worldsense and Euro-Christian (British) worldview needs to be taken seriously. In no way am I advocating a dichotomized construction of these cosmologies. Rather, I highlight the stratified social relations of power between one, a worldview, and the other, a worldsense. Also, in no way am I working with essentialist ideas of cultural identity, purity, or authenticity, given that cultural dynamism reflects the realities of centuries of interaction that occur both within and beyond distinct cultures in human social life (Clifford 1988; Greene, 2002; Hall, 2003). Contact, exchange, and the various cultural reconfigurations that are born of interactions between different cultures are part of the human condition. To underscore the gravity of what was displaced in the figure of Esu, the following section explicates Yoruba cosmology and the role of Esu within the Yoruba metaphysical system.

Yoruba worldsense, complementarity, and interconnectedness are central, where Orun (the spiritual world), Aiye (the physical world of the living human and other beings), and Ile (the earthworld) are all interdependent and cannot exist on their own (Olajubu, 2003; Soyinka, 1976). African cosmologies are also metaparadigms that function as maps that guide and direct a people on how to live and exist within their culture (James, 1993: 32). These cosmologies tend to be circular, emphasizing and symbolizing the important philosophy of balance, continuity, and community among the ancestors, the living, and unborn. All are important threads that come together to make the circle whole where in essence, there is no beginning or end, but rather a powerful continuity of life through transmutation.

The circle reflects how the individual exists within the context of the larger community and how the living self is intertwined with the world of the ancestors and the unborn. The circle is both literally and figuratively a symbol of eternity in its continuity. The self, then, is the materialized manifestation of the circle where each is a dialectic mirroring of the other. In this regard, the self is a layering of many synthesized selves that cross over, move in, move out, and move between the spiritual and material realms, making human existence extend throughout the cosmos as opposed to simply being restricted to the world of materiality. What happens to these selves when central figures in the Yoruba worldsense are displaced? How are balance, community, and spiritual literacy affected and how has colonialism re-drawn the Yoruba cosmological map? The figure of Esu gives us some clues as to how this map has been disrupted and redrawn, particularly with respect to the spiritual blockages that are created when this deity who stands at the nexus of the spiritual and material (*Orun*, *Aiye and Ile*) is appropriated, transposed, and defamed into the devil.

Cosmological Difference: Esu and the Yoruba Concept of Ori

One crucial element of Yoruba Indigenous identity mentioned repeatedly by many Yoruba community members was the multilayered concept of Ori, what is literally known as one's head, yet figuratively and spiritually known as destiny, purpose, or calling. This idea was discussed in various ways by community members who were familiar with the complexities of this idea and indicated multilayered understandings. Others offered more literal definitions pertaining to personal experiences. Nevertheless, all Yoruba elders and youth understood Ori to be a popular and important idea regularly used by Yoruba peoples in a variety of ways. However, with the exception of two people, no one mentioned Esu, nor the importance of sacrifice and appeasing deities and spiritual forces in connection with having an aligned destiny. Esu was noticeably absent from this portion of the interviews:

> **Mrs. Oladiran:** Ori is head. But if you look at it the other way, ori is our destiny like God. Like our God, ori is very, very important. The way we call it, ori–Ori is just like [the] head. Our entire head is just God, our destiny.
>
> **Mr. Oladiran:** Okay, you're talking about your ori, you're talking about your destiny. And eh, destiny … once destiny … we are making a mistake when we say our destiny is in our hands. Our destiny is not in our hands, you know why? Because even before you were created, God knows what you are going to become. Our destiny is in the hands of God. Or the only thing we human beings have to do is we just have to cooperate with God and whatever he has laid down for our life will come to pass, that is IF we work according to his precept or according to his plan.

Both Mr. and Mrs. Oladiran anchor Ori in the spiritual realm and in the hands of what seems to be a very Christian oriented God. For them, God is the driving force in their lives guiding what they are meant to do or accomplish while on earth. However, Dele offers a more complicated and less linear understanding of Ori where "good" and "bad" luck are spiritual energies susceptible to human influence and manipulation.

> For Yorubas, there's something about knocking on your head is bad luck kind of thing. Like if a stranger knocks you on the head or something it's back luck or something or that it's not good to get knocked on the head. Or you know when parents threaten you when you're behaving bad they say, "*ma fun e nko.*" Another thing is when I was coming to Canada, my Grandma, she did something to my head … she called some man that did something on my head as if [for] some kind of protection.

Dele's discussion reminds us of Yoruba worldsense and the constant awareness of various spiritual forces that can assist or hinder a person. His discussion also suggests ritual and ceremony as vital tools used to empower individuals while simultaneously ensuring that connections with other community members remain intact. Similar to Dele, Mrs. Olusanmi understands Ori to be the merging of the material and spiritual, where spiritual energies can be recognized as the driving force behind what appears to only be physical:

> *Ori* is more or less what you call "*Eleda*" [The Creator]. Well, it's destiny too. *Ori* is considered Eleda, which is your creator and the head. People believe in it ... With some people, they believe that if anybody [has] offended them, once they beat the head like this [knocking on her forehead with an open palm], that, whatever they say, it's going to happen to the person. Let's say I have such a power and somebody [has] offended me ... and I say, oh my goodness you did this to me, then I beat my head like this [knocking on her forehead with an open palm again], in the power of my head, this is going to happen to him or her. Us Yoruba believe that that thing will happen, and it does.

Mr. Awoniyi's discussion of Ori emphasizes its multilayered quality and the various beliefs that Yoruba peoples hold. In addition to understanding Ori as the fusion of spirit and matter that is interdependent, he offers a detailed understanding of destiny as a state of being that is folded into the circumstances surrounding one's birth and identity. Mr. Awoniyi also notes the significance of rituals and ceremonies in determining a child's calling to ensure that a good destiny is properly aligned for the child and that it remains intact:

> Okay, now Ori, like here in the human sense we see it as the physical head, but it's not just the physical head. In English yes, head is head. The only time we change head in the English thing is when we say, yeah he is the head of an organization, and in Yoruba we say he's the Olori. Like, we will talk of the government, oh, *Olori ijoba*. But ori in Yoruba is not just ori, it's like an aura that surrounds you as a being and it's not just a one layer aura, it is in stages, in levels ... Now all those [that] are referred to as Ori, ori could also mean crown. Because it depends on how you see it, how you describe it. Ori is also your destiny and we believe that it is something you bring with you from *Orun* [heaven]. Now there are people that are born in the Yoruba culture, they are born with a certain type of thing that comes with them. Like, like the *Ojo* I think comes with a cord around his neck, you know the umbilical cord is around the head. So we say, "*Ori ti ojo gbe wa aiye niye.*" [that is the destiny that this child with the cord around his neck brought with him]. When a child is born with the cord around the neck, we call him *Ojo*. Dada comes with locks in their hair, you understand? Dada [is] a group of people that may not be connected physically, but spiritually, unknowing, they are connected, you understand? So you call all of them together ... *Ori ti Dada gbe wa aiye niyen* [that is the destiny that Dada people with prelocked hair from heaven brought with them]. And um, You don't cut it [their hair]. You just leave it [and] it grows and all you do is just keep it clean, and it becomes a natural lock. Locks just like you have. There is nothing you do to it. Then it gets to a certain age when you want to really cut the hair. For some of them [Dada people] if as soon as you see the locks you cut it, they might react to it; get ill, create fever or they become feverish. You have to really consult some oracle or do some things, you know to avert

> anything happening to that child. So, when it gets to a certain age–. Um, from one example that I noticed which was my sister, they had to um, it's like we call it saara. Which is just a type of sacrifice and calling all little children in your whole area and providing them with lots of goodies, you understand? It's just making it subtle for that child ... that now you are cutting the hair. And it will be in the memory of that child forever that you know what? "The day my hair was cut, oh boy did I have a very big party!" you understand? But mind you, it's not just a big party, because you do not know what ori [destiny] that each of these child[ren] that you have called ... has. So now, you are appealing, you are giving them gifts, that you know, accept my child into this clan of children, you understand? And then you have really ... I will say it in a very shrew or crude way, it's like buying your way into the community. So, okay, now we are now going into different levels and different layers ... but I will just break it down to just the aura, the layers of aura that surrounds a human being.

Mr. Awoniyi's discussion of "Saara" (a type of community-based sacrifice) reminds us of the centrality of community in Yoruba culture especially for ushering in important life events like birth. Within Yoruba worldsense, birth is a journey of travel from pure spirit to earth, where one brings certain aspects of one's being with them that not only make them unique but also give some clues as to what the new addition to the community's destiny is. Hence the widespread and popular phrase "*ori to gbe wa*": "the destiny that he or she brought with them." Similar to Mr. Awoniyi, Sade defines Ori as a multilayered state of being where the spiritual cannot be separated from the material. She emphasizes the significance of sacrifice and ritual where an individual's purpose or calling is concerned:

> When you talk of ori, it's just as [any] ordinary human being [where] you have your head, you have your head. Your brain is there, everything is there. But then when we [Yorubas] talk of destiny, the belief is that before you come into the world you have chosen whatever you are going to be. That's our own belief, I don't know that of the Western world but then even sometimes in the olden days they used to go and consult the oracle and find out the type of destiny that that child brought into the world. So they try to guide to continue with that ... and if there is anything that is not too good they see—they ask what can they do in order to help him move a better way, get a better destiny, right. Yeah, they used to do that. So destiny can never be changed they say, but then if it's bad destiny, they still have to appease God or do something in order for that bad destiny to change.

Here, I suggest that Sade's discussion of alignment of Ori is twofold. She speaks about alignment of the child with his or her destiny, and what is at the core of this discussion is the importance of an individual's alignment of their body, mind, and spirit/soul, what is popularly spoken of as the body, mind, and spirit connection.

In short, Ori is a salient element of Yoruba worldsense because it maps out how people know and understand their world. George Dei (2000) notes that Indigenous knowledges are grounded in the cosmos, are holistic, experientially based dynamic knowledges in which "through the process of learning the old, new knowledge is discovered" (6). As discussed by many of the Yoruba community members interviewed, there is no separation between the secular and the sacred in Yoruba spirituality because spirit is embedded in everyday life from the time before one is conceived, through their birth, throughout one's physical life as a human being, and after one dies (Mbiti, 1975: 2). As discussed by Mr. Awoniyi, Sade, and her son, Dele, ritual is one of the primary sites for interaction between the spiritual and physical (Olajubu, 2003: 3). Ritual helps to maintain links to ensure that connection with the unseen remains active and unbroken. An important element of Yoruba rituals and ceremonies is the communication between the living and metaphysical beings, and how these beings are recognized as vital members of the community. Yoruba rituals are often carried out to mark important life-altering stages such as birth, baby-naming ceremonies, initiation, rites of passage into adulthood, marriage, and burials. All of these life stages are infused with Indigenous Yoruba spiritual beliefs as a means of communication with all members of the Yoruba community. The Orisa that is responsible for ensuring that these links remain intact is Esu, particularly where offerings and sacrifice during the performance of these rituals is concerned (Abimbola, 1976). Oyeronke Olajubu (2003) describes Esu as key in performing the ritual of aligning an individual with their Ori:

> Ori (head) is conceived by the Yoruba as a representation of the inner essence in humans; it symbolizes the individual's essential nature. It's conception is underscored by the Yoruba perception of self as interior and exterior the latter being dependent on the former … The individual whose Ori is to be propitiated sits on a mat, dressed elegantly. Items prescribed for the occasion by Ifa divination are each placed temporarily on the participant's head, one after the other. Prayers are offered for the participants to solicit the support and guidance of Ori. Thereafter, a sacrifice is offered to Esu (god of the crossroad and messenger of the deities), to ensure free passage for the requests made. (33)

Rowland Abiodun (1994) builds on the concepts of interior and exterior self, also known as inner and outer head:

> In the visual arts, notably in sculpture, ori-ode ("physical head") is the focus of much ritualistic, artistic, and aesthetic activity. Not infrequently, the head is given a place of visual command by proportionally subordinating all other parts of the body to it. The enlarged head is further emphasized by detailed artistic treatment with elaborate

> coiffures, crowns, or other headgear ... the absence of ori and oju in any sacred or secular activity, whether artistic or not, would be tantamount to anarchy in the human and spiritual realms of existence. There would be no *ase*. (77)

As messenger and deity of the crossroads between the material and physical, Esu has the charge of traveling, translating, and judging whether the offered sacrifice is appropriate for the designated spiritual cognates. Esu carries the concept of interior and exterior head, and it is this deity's role in the dualistic dimension of this model that I focus on. This notion is also of importance because it mirrors Esu's role and function in Yoruba worldsense, and is important precisely because it entails both literally and figuratively crossing or spanning over more than one aspect of a particular entity. In this case, it is Ori: a state of being that, as many of the Yoruba members interviewed assert, cannot be minimized. The sense of being dual, or multi-layered (parallel to Ori), is of particular relevance to Esu because it signals how this Orisa cannot be conflated and confined to a singular role, function, or identity. Nevertheless, this is exactly what occurs when Esu is re-made as Satan. The duality of this Orisa (encompassing material and physical, left and right, male and female, hot and cold, red and black) is not parallel to dichotomized qualities that exist, never touching, yet stratified side by side. Rather, Esu's dualities are *intersecting* ones that operate in concert with one another, hence the association of this Orisa with the crossroads and its particular insignia as one of flux, transformation, and movement. These dualities that Esu is charged with are also complementary—a key principle in Yoruba philosophy and cosmology (Abimbola, 1976; Oyewumi, 1997). In this way, Esu represents the realm of possibilities and multiple happenings that both incorporate and extend beyond dualisms. The crossroads entails an image of this Orisa that unites and divides in the process of attaining what Eliade would call "divine totality," which is defined as incarnate perfection (Lawuyi, 1986: 307).

Esu's role as the judge who either blocks or "ensures the clear passage" of requests made from the material to the spiritual realm is also noteworthy, solidifying this Orisa's fundamentally important role that operates in keeping the "lines of communication" open and linked between the physical and metaphysical. In this way, Esu is also charged with the task of ensuring that people are properly aligned with their destiny. That is, assuring that their *Ori-inu* (inner metaphysical head) and *Ori-ode* (physical outer-head) are aligned. Given this, Esu is a key figure in ensuring that Indigenous Yoruba understandings of the self (popularly known as the body, mind, and spirit connection) are in sync. Importantly, Esu is also endowed with *Ase*, which is a vital life-force and divine essence that imbues sound, space, and matter with energy to restructure existence to transform and control the physical world (Abiodun, 1994: 78).

Another dimension of Esu is this Orisa's role as trickster or mischief-maker who can cajole one into confusion and into possibly making regrettable decisions (Idowu, 1996: 80).[3] However, Esu tempts people in complex situations by offering a number of choices to see how they will handle such a situation. Esu is that part of the Divine that tests and tries people to reveal or see what the true nature of their character is, hence the application of the trickster title (Awolalu, 1979: 28). Bibi Bakare-Yusuf and Jeremy Weate (2005) elaborate on the complexities of Esu as trickster:

> [Esu] challenges us to reflect constantly on our lives and not get too blinded by habit. He is cocky and masterful, but against cockiness and mastery. At first sign of complacency, Esu keeps us in check by introducing chaos and confusion … He is sometimes referred to as the "devil." This is not because he is spiteful or the devil, as the Christian mistranslation of his characteristics would have us believe. Rather, he wants us to always be alert, vigilant, and to make active choices by questioning our sense of certainty and unexamined faith in the world. (331)

Communication, translation, sanctioning, and bearing sacrifices from the material to the spiritual, trickster, character builder, deity of choice, and of the crossroads include some of the many facets and faces of this complex Orisa. It is the complexity and multilayered role and function of this deity that made me pay attention to the contrasts between how Esu was constructed by Yoruba community members, and how this deity is understood in Yoruba culture according to the Indigenous roles and responsibilities of this Orisa. The continued evocation of Esu as the devil among the Yoruba elders and youth was of significance, and therefore deserving of deeper exploration. Because Yoruba worldsense is rooted in holistic harmony between its various dimensions and entities, and because Esu is a key figure in the stability and continuity of this harmony, to misunderstand or disrupt this Orisa is to disrupt a key element that endeavors to ensure the balance of material and spiritual harmony that is crucial to the Yoruba cosmos. The reduction of Esu's significance to a satanic figure is an example of blockage to critical spiritual literacy that is anchored in cosmological imperialism and the inequitable denial of difference among cultures. In the Yoruba context, this is when the varied and complex layers of Esu in Yoruba cosmology are stripped, reduced, and reconceived into Euro-Christian conceptions as the lord of evil and antithesis to God. Historically, this re-casting and transposition of Esu has occurred primarily through missionary institutions and proselytization.

Cosmological Imperialism: Biblical Hegemony and the Re-constitution of Esu as "the Devil"

Alongside Yoruba elders and community members' repeated reference to Esu and other Yoruba Orisa as idols[4] was the insistence that Esu was also the devil. This can be summed up in Mrs. Oladiran's opinion that "this idol [Esu] only knows about harm and how to hurt people because he is the devil." When I asked community members to share their thoughts on traditional or Indigenous Yoruba spirituality, it was suggested by Mr. and Mrs. Oladiran, Mrs. Olusanmi, and Mrs. Awoniyi that this was a spirituality that was practiced during "*igba imo*": translated as a time before the advent of Christianity where primitivism, ignorance, and not knowing God reigned. Such a perception indicates an internalization of the widespread construction of Yoruba worldsense and Indigenous ways of knowing God as primitive and uncivilized. This internalization of colonial ideology warps an individual's ability to critically read and engage with the sacred, and is especially true when considering Esu's key role in Indigenous Yoruba spirituality and how this figure has been transposed unto the devil. The association of Esu with Euro-dominant notions of evil curtails the desire of Yoruba individuals to openly embrace the everyday experiences and practices that are guided by this deity and other Indigenous ways of knowing.

One of the most powerful tools used to block spiritual literacy and engagements with the sacred in Indigenous contexts has been the missionary usage of the Bible. Historically, the Bible is a text that has combined Euro-Christian worldview with Eurocentric notions of literacy for dissemination among Indigenous peoples. This was done for purposes of religious conversion and socio-economic subjugation. The necessity of engagement with the Bible to become literate and educated was not merely a religious enterprise but also a political, social, and economic act that lay the groundwork for colonization. Ologunde (1982) reminds us of the pivotal role the Yoruba language played as vehicle in conversion of the Yoruba to Christianity. He cites missionary control as an era where the central goal was to civilize, enlighten, and spread the word of Christ through the Bible. During this time, missionaries learned Yoruba, developed a Yoruba orthography, recruited and taught this orthography to Indigenous Yorubas so that they would become many of the teachers and catechists through which the gospel could be spread. On the heels of the production of a Yoruba orthography came translation of the Bible into its Yoruba version. In this era it becomes clear how Eurocentric notions of literacy were inextricably bound up with Christianity and the missionary agenda to have it spread among Indigenous communities. And this was accomplished primarily through translation of the English Bible into Indigenous languages. While technically the Yoruba language was being developed in the sense that it was translated into an orthography in the form of an alphabet, this was done solely for purposes of domination through conversion, and coded as such in colonial discourses of enlightenment, progress, and civilizing projects. In the translation of an oral culture into an orthography, one form of literacy replaces the other as normative. In other words, Eurocentric notions of literacy are positioned as the norm while Yoruba Indigenous literacies are condemned as an inferior and obsolete era of not knowing God. Here, I note that in no way am I advocating a dichotomized construction of literacy because the two forms of reading are (orthographic and sense based) complexly interwoven and do overlap. What I highlight is the manner in which one's respectability as an orthographically literate and educated person is entangled with being a Christian versed in biblical knowledge and discourse. Said differently, the Bible and being able to read this text were central to one's social status as civilized and were the building blocks for this type of prestige.

Fascinatingly, of the more than 400 Orisa, Esu is *the only deity* that "makes it" into the translated Yoruba Bible. This is not a random occurrence. Rather, I argue that this reflects a conscious missionary effort to sabotage and convert Esu out of its unique role and function within the Yoruba metaphysical system. Reassigning Esu the role of the devil in the Yoruba Bible not only meant stripping this Orisa of its Indigenous responsibilities and powers, but it also suggests that the Yoruba Bible itself emerged as an amputative imperial tool because Esu the Orisa is replaced with Esu the devil. A number of Yoruba elders and community members in this project echoed this sentiment when they repeatedly referred to the Bible as their main source of evidence in proving Esu to be the devil and essence of evil. When I asked Mrs. Awoniyi to elaborate on her position that Esu was the devil and how she came to know this, she passionately replied:

> That is what they call it!!! Esu … Okay. *Je kin le mo 'le fun e*! [translation: Let me break it down for you now]. They say that Esu has red and black, that it follows harm and that if it really wants to do harm, it follows the person until it does this harm.

> So, Esu ... even in the Bible they say he goes up and down throughout the universe until he achieves what he wants to do. So in that case, Esu is an evil thing ... an idol that is the devil.

Interestingly, her partner Mr. Awoniyi understood Esu to be an Orisa that was not the devil. He argued that the complexity of the Yoruba language did not make for exact or accurate English translations. He also contended that Esu's name was used in multiple ways by Yoruba people to mean different things:

> Esu is not ... you see Yoruba is complex, it's a very, very complex language. We cannot literally say that Esu is the devil. Ah, you can't just give it a direct meaning, to say Esu is the devil; just to get away from a long conversation you just say Esu is the devil for the person who speaks English to understand where you are going. But this is not so to the fullest extent because ah, ah even in the Bible, Satan is not described as the Devil. In the Oxford dictionary, it's where the devil is related as Satan, but in the Bible, Satan is Satan and Esu is as powerful as being somebody who is satanic. You see, with this Esu, it's the way you say it! You could mean Esu to be really, really, really bad and you could use Esu just for conversation to say that a person is giving you problems. Whereas Esu is the extreme of all evil, which is Satanic.

Mr. Awoniyi's position appears somewhat contradictory in that while refusing to conceptualize Esu as the devil, he likens this deity to being "really bad," "satanic," and problem ridden. However, there is a powerful counterhegemonic subtext in Mr. Awoniyi's position in the sense that he insists on multiple, and therefore more complex, readings of Indigenous Yoruba spirituality that must be anchored in equally sophisticated understandings of the complexity of the Yoruba language. Significantly, Mr. Awoniyi recognizes that the cultural meanings, ideologies, and philosophies embedded in Yoruba language and worldsense can and do become obscured in English and Euro-colonial translation. This is particularly so in colonizing contexts where social relations of power are inequitable and racist Eurocentric constructions of Yoruba Indigenous ways of knowing are situated as the norm. Mr. Awoniyi's position echoes Thiong'o's (1986) and Oyewumi's (1997) argument that Western knowledge production processes are culturally specific to the West and that Indigenous African languages should be taken seriously on their own terms in scholarly research. Given this, Mr. Awoniyi's insistence on multiplicity reflects an assertion that knowledge be anchored in a cosmological context. Here, Mr. Awoniyi alludes both to the complexity of the Orisa Esu and his agency as a Yoruba elder in this study. Mr. Awoniyi's understanding is in stark contrast to the one-dimensional Eurocentric constructions of Esu voiced by other Yoruba elders and youth in this project, many of whom repeatedly referenced the Bible as an authoritative and reliable source for their position that Esu was, in fact, the devil. In response to my question about what spirituality means to her, Yinka passionately answered:

> In the way that I was raised, being a Christian I find is a powerful thing. And the Bible basically guides everything that I do and believe. That's the center of the whole world is my religion ... Being a Pentecostal compared to Catholic or Anglican—'cause I've been an Anglican before—the way in which they worship is completely different. What I'm saying is Pentecostals—well I'm not gonna speak for everybody else but, the Pentecostals that I know have the Holy Spirit and that's what we work with is the holy spirit and it says so in the Bible. So I can't speak for someone that's Catholic, someone that's Baptist and someone that's Presbyterian but I feel because I'm Pentecostal, I have a sense of power because I have the holy spirit with me.

Similar to Yinka, her sister Tunmi also places great emphasis on the Bible as a sacred text with divine authority. In response to my question on whether she felt there was a difference between spirituality and religion, she remarked:

> Well there are a lot of people that are not religious but they're spiritual. They have a relationship with God and they don't really connect themselves to any certain religion ... I guess when I was younger I really didn't have much of an appreciation for my religion. Um, but as I got older I could see like, you know what I mean, like I could see my parents' values and the things that they hold dear to them, like reading the Bible and praying. And I even sometimes see it in me whenever I'm doing stuff so it's like, you know what I mean. Like it's something that you see growing up and either you respect it or you don't. But I get a lot of answers when I'm confused about something. The Bible guides me and I never used to see it but now, it's really important, an important source of spiritual guidance, like you know what I mean?

The discussion here is not to deny participants their agency. Nor is it to deny the Bible as a sacred text that interviewees have a right to value and hold in high spiritual regard. What I want to highlight is how the Bible and biblical discourse have been used as hegemonic tools in colonial and imperial projects, especially when this text is touted as the universal and sole source of spiritual knowledge at the expense of Indigenous spiritual knowledges.

As an unquestionable source of divine authority, biblical discourse also becomes unfailingly dependable. Without a doubt, one of the most notorious figures in Biblical discourse is the devil, otherwise known as Satan. The Bible is not the same without the devil, for he is needed as the antithesis to all that is good and all that is God. Esu is a central figure in the Yoruba cosmos, while Satan is a central figure in Euro-Christian worldview. However, this is where the similarities between these two figures end. In the Yoruba Bible (and largely in Yoruba Christian social life) the figure of Esu is stripped of its complexity in Yoruba worldsense and conflated into a simplified singular role, function, and identity: the garb of consummate sin and evil Satan and the devil. In the New King James Version of the English Bible, the title "devil" appears 35 times (Youngblood, 1995: 352, 1131–1132). In all 35 of these instances, my research shows that Esu is repeatedly reconstituted, bearing the title of "the devil" in the Yoruba Bible. This continuous displacement is particularly striking in the Book of Revelations. Below are three passages that demonstrate this misguided transposition from the King James Version of the Bible (1989) to the Yoruba Bible, *Bibeli Mimo* (1960, 2nd ed.), which was first translated into Yoruba by native agent and missionary Samuel Ajayi Crowther:

Rev. 12:12
Therefore rejoice, ye heavens, and ye that dwell in them. Woe to the inhabiters of the earth and of the sea! For the devil is come down unto you, having great wrath, because he knoweth that he hath but a short time (178).

Ifihan 12:12
Nitorina e ma yo, enyin orun, ati enyin ti ngbe inu won. Egbe ni fun aiye ati fun okun! Nitori Esu sokale to nyin wa ni ibinu nla, nitori o mo pe igba kukuru sa li on ni (1057).

Rev. 18:2
And he cried mightily with a strong voice, saying Babylon the great is fallen, is fallen, and is become the habitation of devils, and the hold of every foul spirit, and a cage of every unclean and hateful bird (180).

Ifihan 18:2
O si kigbe li ohun rara, wipe, Babiloni nla subu, o subu, o si di ibujoko awon emi Esu, ati iho emi aimo gbogbo, ati ile eiye aimo gbogbo, ati ti eiye irira (1060).

Rev. 20:10
And the devil that deceived them was cast into the lake of fire and brimstone, where the beast and the false prophet are, and shall be tormented day and night forever and ever (182).

Ifihan 20:10
A si wo Esu ti o tan won je losinu adagun ina ati sulfuru, nibiti eranko ati woli eke ni gbe wa, a o si ma da won loro t'osan-t'oru lai ati lailai (1062–1063).

However, while the title of the devil is consistently displaced unto Esu, this is not the case for the name Satan. In other words, although Satan is largely translated as Esu in the Yoruba Bible, there are a number of instances (approximately 25% of the time) where a Yorubacized version of Satan appears as "Satani" as opposed to Esu. This lead me to question why. Why did the translator, Samuel Ajayi Crowther, not completely displace Esu by always casting this deity as Satan in the same way this was done with the title of the devil?

Although Satan and the devil are often thought of as one and the same, it could be that Crowther wanted to distinguish between Esu as Satan and Esu as the devil by highlighting the difference between the two. And in doing so, he would be demonstrating his unfailing commitment to Christianity. Consider the fact that "the devil" is a title that connotes pure evil, while "Satan" is the name of a being or fallen angel that has a history of divine origin from God and heaven. As Lucifer, Satan was once good and noble. However, the devil does not share this paradox of having noble yet savage qualities and tends to be understood as evil incarnate. Esu as the devil suggests unwavering evil while Esu as Satan suggests something different. Interestingly, if it is Satan that has some noble qualities and the devil that does not, it then seems that Esu still remains an absolutely subjugated manifestation of all that is evil, the antithesis to good and inherently demonic. On the other hand, it is possible to read Crowther's absentization and failure to name Esu in some parts of the Yoruba Bible (be it as Satan or the devil) as a type of subconscious, counter-hegemonic resistance to the demonization of this Orisa by retaining Esu's role and function in Yoruba cosmology *precisely through this absence*. It also highlights that while the demonization of Esu is undoubtedly symbolic of the political, social, and religious agenda in the European colonial project, this project was never complete. Therefore, this highlights the power of resistance and the power of Esu the Orisa to elusively maneuver and slip out of a total displacement from its Indigenous context and role. It is impossible to pinpoint exactly what Crowther had in mind when using the name Esu to index the devil. However, judging from Crowther's explicit positioning of Yorubaland "as a land of heathenism, superstition and vice" (Ajayi, 2001: 29), it is clear that his conception of Esu is not affirming. It is also one that does not demonstrate but, rather, elides a thorough understanding of Esu's multifaceted roles and responsibilities, be it intentionally or unintentionally. Paradoxically, despite Crowther's conversion and declaration of absolute commitment to Christianity, he also carried the cultural burden of blockage to Yoruba worldsense in his translation of Esu. Hence, Crowther himself can also be read as a metaphor for cosmological imperialism and spirituality as a site of contestation.

(Resisting) Cosmological Imperialism: Racism, Gender, and Black Feminist Readings of Evil

Whereas most community members vehemently defended their position that Esu was the Yoruba equivalent of the devil, during this portion of the discussion Bisi, Sade, and Mrs. Fayemi cited racism, not Esu, as evil and preferred to spend more time discussing how racism affects their lives. This emphasis on racist oppression highlights their agency and what Chezia Thompson-Cager identifies as the Black feminist tradition of "naming as power" (cited in Collins, 1990: 111).

As a form of oppression that denies Blacks and other racialized peoples their humanity and minimal access to jobs, housing, education, and other social and material rights (Dei, 1996; Henry et al., 1998), racism is traumatic and often debilitating. This reality parallels community members' experiences of racism that they shared during the discussion of Esu and evil:

> **Bisi:** Like I said when I'm praying I don't talk about the devil or all these evil things right; I talk about self-development, to me anyhow. So, I take that everyday … I don't even … like … the racism thing, unless you're slapping me in my face and calling me a Black whatever, that's when I'll let it really affect me. But I know that I'm gonna do whatever I can to show you that this is what it is. Like, I'm gonna get that job, or that A+. And not even to show you, but to prove to myself that I can be this and do this, so … to me, that's what evil really is, but I don't let it stop me.

Bisi notes that for her, spirituality does not involve engaging in the negative and dismisses the value of an explicit discussion about the devil or anything that she characterizes as evil. Instead, Bisi takes a proactive approach where she focuses on developing and nurturing her being in more affirming ways because this is what allows her to continue succeeding in the Eurocentric context of Canada. This was Bisi's way of proving racism and racists wrong in the hegemonic scripts that have been discursively constructed about her racialized self and being.

Similarly, Sade offers compelling evidence of how much more violent and harmful experiences of racism are than Esu could ever be. Her discussion is a powerful example of her agency and resistance to dominant hegemonic constructions of Indigenous Yoruba culture, as well as to explicit forms of racist oppression that she is forced to negotiate on a daily basis:

> I think [that] just because it's [Esu] in the Bible that the devil is Esu, that is why people believe this. But it's the wrong notion of him … Everything we blame on the devil, but you know, Esu isn't the reason I have problems here [pointing outside to the snow, making reference to Canada]. You know, I was in the subway one day and one lady asked me how is life in the jungle when I first came to Canada. And I just joked about it, because I know about the zoo and you know me I'm a joker—I said, "Well, sorry, I've been to the zoo" and I said everything in the zoo in Nigeria has been taken to the zoo in Scarborough [a city in the greater Toronto area], so they're all like that … [laughing]. And people were looking at her laughing and she looked stupid. And there was a time I got a job in a company as a data entry clerk, and the data entry people were like four in the office. They relieve the receptionist. So every time I went there, they [would] put the phone automatically on right to the answering service and I'm like, "why do they do that?" But if people come in, I talk to them. But when they call, then the phone goes on to answering service!? So I asked my manager, "what happened? Why is it that everybody is there and they answer the phone, but why don't they give it to me?" And she said oh, it's not her, but it's the supervisor who is more senior than her … They say that because, maybe they won't understand my accent. So I was writing one of my essays in York [University] one time and this was one of the examples that I used in my essay when we were doing racism in Canada. So, when I finished my essay I showed my manager and she says "you know it's not me" and I said "I know it's not you" and she's a Canadian too. So there is racism of course and I've experienced it … That is the kind of stuff that harms me, not Esu as the devil. As I said before, when you speak of Esu as devil I know it's not true. People have the wrong idea about it. I have more problems because of [my] skin color than I do because of an idol.

Here, Sade claims and utilizes her power and agency by renaming racism as evil instead of the dominant perception of Esu as such, knowing that racism has oppressed her and negatively affected her life more than the Orisa Esu ever could. Patricia Hill Collins (1990) discusses the Black feminist tradition of renaming as a form of re-articulation where Black women's independent self-definitions empower us to change the conditions of our lives.

Although Mrs. Fayemi does believe Esu to be the devil, similar to Bisi and Sade, she also identifies both racism and sexism as huge barriers to her success in Canada in terms of access to jobs and education:

> What is Esu? Esu is the devil. Anything evil really … You know, my understanding of evil [is] somebody who do[es] harmful things to a human being; somebody who

> doesn't wish somebody else good things ... You know what, I always tell my children that, "One thing you have to know, even though you are Canadian born, you are Black. They will look at you as being Black first before looking at you as Canadian." I always tell them that, don't get so comfortable that "oh, because I don't have accent, I have [a] Canadian birth certificate, that people will take you as the way they take White children or White kids," no! They look at you as Black before they look at you as Canadian. And once they see your name is different—Especially at work, once they see your name, they know that English is not for you no matter how good [the] English you write [is]. You know, they believe that English is not for African people. And they use your name to mark you. At work, oh! They look at you; we are just like second-class citizens. No matter how smart you are, because of your color, our problem is you are Black and you are a woman; we have two absurdities.

Mrs. Fayemi offers a powerful critique of Canadian racism and White supremacy by shifting the paradigm of how evil is discussed from one that positions the Indigenous figure of Esu as evil, to one that instead situates colonial and White supremacist discourse and practice as such. Her discussion also reinserts gender and sexism into the discussion. Mrs. Fayemi's insight and reference to racism and sexism as "two absurdities" clearly demonstrates her Afrocentric feminist consciousness that is grounded in a rejection and re-articulation of Eurocentric masculinist discourses (Collins, 1990). This also provides an important segue into gender, Esu, and how this Orisa's transposition as the devil includes the hegemonic construction as demonically male/masculine figure in Euro-Christian discourse.

Bible-Thumping Imperialism: Gender and the Essentialized Masculinization of Esu

Esu becomes increasingly important in this study when almost all Yoruba community members' constructions of this Orisa were taken up as distinctly male. This affirms what Oyewumi (1997) calls the naturalization of Western gender discourses in African thought, life, and experience. It is important that community members' uncritical identification of Esu as an exclusively masculine figure be given serious consideration, namely because this brings to light the dominance of Western constructions of gender as the norm. When I pointed out to Yoruba elders and youth who were interviewed that while the devil might be male, the gender of Esu as an Orisa was not quite as clear-cut or one-dimensional, many responded by citing the Bible as proof that Esu was both the devil and male. Mrs. Awoniyi argued that "the Bible does not lie" and cited the story of Lucifer, God's fallen angel, as evidence that the devil is male, as is Esu. For her, the issue was not Esu's gender but that this Orisa was simply evil and the devil. Similarly, Mr. Fayemi argued that Esu was male because only men could have that kind of power. He had the following to say when asked to elaborate on his understanding of Esu and male power:

> Yeah, Esu is the devil. Yeah, Esu, that's the meaning: it's the Yoruba word for the devil or Satan. And he is very powerful you know, because evil is powerful and you know that he fought God. He still fights God and he knows how because he used to be his [God's] favorite. It is the kind of war that men do … for power. That is the kind of power that God and the devil have. One is good, while the other is bad, but no matter how powerful the devil is, Esu cannot defeat God because God is the ultimate man power!

Mr. Fayemi spoke of Esu as synonymous with the Christian devil, who is undoubtedly male in biblical text and Euro-Christian discourse. What is also of interest is Mr. Fayemi's construction of power as resoundingly male. His position reflects Euro-Christian masculinist views that eclipse Indigenous understandings of power, gender, and the complexity of Esu as an Orisa. Constructions of gender as a binary are also emblematic of cosmological imperialism because the dominance of Eurocentric worldview does not allow for Indigenous understandings of gender to exist as valid ways of understanding one's culture and spirituality. With specific reference to gender, cosmological imperialism echoes Oyewumi's (1997) position that "gender does not operate in Yorubaland [many Indigenous contexts] in the same way it does in Western discourse" (x, xi). The gendered mistranslation of Esu is what Oyewumi calls "masculinizing the Orisa" where, while some Orisa are thought of in gendered terms (i.e., Oya as female and Sango as male) others are either gender neutral or their gender is context dependent (136). That is, their gender is dependent on the worshippers themselves and how they imagine and engage with that particular deity (Oyewumi, 136: 140–142). Esu is one case in point.

In her analysis of an earlier study conducted on Yoruba devotees of Esu, Ayodele Ogundipe concludes that Esu's depiction in Yoruba spirituality is both male and female (cited in Oyewumi, 1997: 173). However, Oyewumi continues the discussion by cautioning that Esu's sex or gender should not be interpreted as one of ambiguity or androgyny, but rather as genderlessness. She argues that Esu's gender is incidental to this Orisa's role and function in the Yoruba worldsense and underlines the vital fact that ori (metaphysical head/destiny) is without a sex or gender (173–174). The latter is a crucial point. I go back to my earlier discussion of Esu's role and function in Yoruba worldsense with specific reference to Ori and how essential Esu is to an individual's proper alignment with their destiny. Again, I propose that this Orisa's complexity (how Esu is multilayered with the insignia of the crossroads, flux, movement, transformation, and role of keeping communication between the spiritual and physical worlds) has nothing to do with gender. Yet, due to the advent of Christian proselytization, Esu has been locked into Eurocentric conceptions of gender and completely masculinized in the Yoruba Bible. This also is in part because the devil and Satan are masculine figures in the English Bible and Euro-Christian worldview. Inevitably, hegemonic discourse is retained and imposed on Esu.

Another element that must be considered in the mistranslation of Esu is that of language. Similar to Mr. Awoniyi, Sade reminds us that the role language plays in the translation process (gendered or otherwise) needs to be taken into account, even where sacred texts such as the Bible are considered. In Sade's words:

> I think ... when you look at it, at Esu ... everything we blame on the devil. When we see something good and bad and we go to do the bad one, and then the consequences come, then we say, oh it's the fault of the devil; it's the devil's doing, instead of taking responsibility as Christians should. I think maybe we make Esu to be the scapegoat of any circumstances that are bad. And in the Bible, they gave the wrong name to Esu because Esu shouldn't be the devil in the Bible. Maybe because of the differences between the English language and the Yoruba language ... language is very important sometimes. There are some words even in our language now that we don't even know how to say it in English, you know. So they gave a wrong notion of Esu because Esu himself is like all those gods [Orisa] like God of thunder, God of water etc. I think [that] just because it's in the Bible that the devil is Esu, that is why people believe this. But it's the wrong notion of him.

Sade is drawing attention to the significance of language and how often there are no sufficient equivalents in English to replace or translate certain words from Indigenous languages. And in this instance, Sade is saying that there is no English equivalent for Esu as a deity and Orisa. Sade's understanding of Esu is more complicated than the binarized constructions of this Orisa as either male or female. Sade understands that the cultural meanings and ideologies embedded in English are limited where Yoruba language, culture, and worldsense are concerned. She also understands that a fluid and genderless (Oyewumi, 1997) Yoruba figure such as Esu becomes fixed in English and can therefore only be spoken of as a "he" or "she" because these are the only two categories that exist in dominant Euro-Christian discourse, and especially the Bible. Yet, Sade navigates these difficult terrains because her consciousness of the philosophical and cultural differences between the Yoruba and English languages allow her to. In this sense, she is able to speak of Esu as male in English due to the rigidity of English and the dominance of the devil in Christian biblical discourse. However, she is also able to retain Esu's fluid gender identity in Yoruba discourse because she knows that there is no equivalent to Esu in English, and because she knows that Esu is not the Euro-Christian devil. Sade is cognizant of the importance of understanding cosmology when translating an idea from one language to another. These sentiments are echoed in Aboriginal scholar Marie Battiste and James Henderson's (2000) work, which encapsulates the idea of cosmological imperialism:

> Imposing foreign categories on other people's lives is sheer folly, primarily because the categories do not apply, at least not without serious revision … The missionaries, for example, attempted to match indigenous spiritual concepts to those of the Christian religion: they sought to exchange one set of verbal symbols for another, based on the assumption that there was a universal God who created the world and a mercantilist version of exchange … The missionaries' purpose consciously or unconsciously, was to use the sounds of indigenous languages to explain the belief forms of Christianity. (80)

As in the earlier tension concerning language and subtitles in the film *Daughters of the Dust*, undoubtedly language is of significance because it shapes ideas about one's culture, philosophy, and identity. Hence, implicit in the idea of Esu as the devil is also the idea that Esu is male: Both are instances where Indigenous spiritual concepts were inadequately matched to those of Christianity (Battiste & Henderson, 2000), resulting in gross mistranslations. In the example of Esu, we are able to appreciate how language, masculinist discourse, gender, Christian ideology, and identity are highly interwoven, particularly where translation and cosmology are concerned.

A related case that provides a powerful illustration of the connection between these social relations is that of Bishop Samuel Ajayi Crowther, the Yoruba missionary who translated the English Bible into Yoruba. Arguing that missionaries and their biblical translation were not innocent, power-barren, nor value-neutral is not new. It has been argued elsewhere that translated alongside the Bible also came Euro-Christian ideas, beliefs, and cultural logics (Oyewumi, 1997; Thiong'o, 1993). My premise here rests with the discussion of cosmological imperialism where the missionary-orchestrated historical moment of biblical translation has also endeavored an ideological and cosmological displacement of key elements in Yoruba worldsense. This is evidenced in the simultaneous appropriation and displacement of Esu in the Bible. This is also marked in essentialized notions of gender concerning this Orisa where Indigenous Yoruba conceptions of masculinity and femininity have become peripheral to Euro-Christian cosmology and the gender constructs embedded in this worldview. This is particularly evident in Samuel Ajayi Crowther's psyche and how he understood his role as missionary to be deeply interlocked with hegemonic Euro-Christian ideas of masculinity. He states:

> about the third year of my liberation from the slavery of man, I was convinced of another worse state of slavery, namely that of sin and Satan. It pleased the Lord to open my heart ... I was admitted into the visible church of Christ here on earth as a soldier to fight manfully under his banner against our spiritual enemies (cited in Ajayi, 2001: 14).[5]

The role of gender in the Indigenous context as well as projects of Christianization is significant here. Clearly, being a man and soldier are essential components to Crowther's declaration of war against his spiritual enemies who undoubtedly are non-Christians. It is the ideology of masculinity and its connection to religious projects that is interesting, particularly as Crowther imagines his mission to be one of war in his self-identification as a soldier. In the Indigenous Yoruba context, soldiers were often male, making it possible that Crowther's utterance may not have been due solely to Western and/or Christian discourses of gender. However, Oyewumi (1997) argues that occupations in Indigenous Yoruba society were also lineage-based. She notes that both women and men alike were and could be soldiers who fought in wars (69).[6] Hence, while one could make the argument that Crowther's declaration of war for Christ was simply due to life as he experienced it prior to being converted, what I underscore is his avowal to patriotically and manfully fight for Christ. This is directly in keeping with dominant Euro-Christian civilizing projects, all of which were patriarchal and gendered. This is also reminiscent of how hypermasculinity is often a vital component in discourses of war where the execution of European imperial projects are concerned. In these times it was first the men who were sent to convert Indigenous peoples in distant lands, and mostly by themselves. If a female figure was ever to appear on the landscape, it was after the male figures had conquered and settled the land. In this context, Crowther's soldier reference alludes to a type of Euro-Christian nationalism because it conjures images of patriotism to Christianity; images that are part of religious fundamentalism. Therefore, the dominant narrative of nationalism here reads as a male-centered and masculine preserve that has a subtext of Christian missionarism. This was especially the case in the early stages of Christianity's insertion into West Africa (Oyewumi 136).

It seems that in Crowther's declaration of himself as a "soldier of God," his enemies also are male. This is evident in the many heated religious debates that Crowther had with Yoruba men in an attempt to convert them away from Islam and their Indigenous faith. Oyewumi (1997) discusses this gendered division with respect to the tensions around polygamy and whether or not a man with more than one wife should be baptized (137). Yoruba men were seen as the principal potential Christian subjects, while Yoruba women entered the debate around polygamy as wives. They had no identity independent from their husbands (137). Hence, in the early stages of evangelization, religious contestation was a masculine concern. This is not to say that African women were not converted and did not contest in multiple and complex ways. However, women were seen as an afterthought to this religious and cosmological encounter.

Similarly, in his classic text on Yoruba culture and society, *The History of the Yorubas: From the Earliest Times to the Beginning of the British Protectorate*, Samuel Johnson (1966) stresses that a Yoruba grammar cannot be produced on the same lines and rules of Latin or English grammar. It is for these reasons that he states Yoruba orthography is still rather defective with "English ideas [being] written in Yoruba words" (xxxiv). This results in confusion of thought and vagueness. Johnson states:

> [I have] on several occasions read portions of Yoruba translations to intelligent but purely uneducated Yoruba men. They would show that they comprehended (not without an effort) what was read to them by putting pertinent questions, but then they would add, "We can understand what you mean when you say … but what you read there is not Yoruba; it may be book language (ede iwe)." … In taking up a Yoruba book one is forcibly struck by the difference in style between quotations of pure Yoruba stories, phrases, or proverbs, and the notes and observations of the writer. The former runs smooth and clear, the latter appears stiff and obscure, because the writer, with his knowledge of the English grammar and language, wrote English ideas and idioms in Yoruba word. (xxxiv; emphasis added)

There are two points of interest in this passage. First, note that African women are absent from the discussion, as Johnson only seemed interested in Yoruba men as intelligent sources of Yoruba language and culture. Again, this is in concert with Oyewumi's (1997) assertion that the introduction of Christianity in Yorubaland was hegemonically gendered and stratified (136). Second, note Johnson's recognition that Crowther's translation of the English Bible into a Yoruba version created a different type of Yoruba, a "book language" version where the formerly fluid Yoruba language becomes stiffened and obscured. It is therefore no surprise that Esu, as the only Orisa that appears in the Yoruba Bible, is also stiffened and oversimplified through essentialist and hegemonic constructions of gender that are endemic to Euro-Christian worldview. Crowther's translation attempted a process where Eurocentric Christian meanings were discursively, textually, and spiritually mapped onto Yoruba worldsense through Esu, demonstrating the relevance of cosmological imperialism. Again, here I highlight the presumption that standard English can effectively and equitably express the ideas, norms, idioms, and symbols of another culture without losing its specificity and uniqueness, particularly where Indigenous societies are concerned. It is no wonder then that many of the Yoruba elders and youth in this book had uncritically accepted and internalized these hegemonic gender discourses to assume that Esu is the devil, and therefore male. Both Samuel Crowther and Samuel Johnson are no exception to this uncritical acceptance of Eurocentric constructions of gender. And with their embrace of Euro-Christian values comes a masculinist position where Yoruba men and male figures are deemed to be the only viable subjects worthy of attention and note, biblically or otherwise.

Conversations of Esu that are informed by Christian worldview are products of benign translatability (Battiste, 2000) and the cultural amputations that transpire when attempting to translate Yoruba concepts into English. This has resulted in Yoruba meanings being lost, while English ideas are gained. The Orisa Esu is a powerful case in point given that this deity was stripped of its Indigenous and counterhegemonic significance in the Yoruba Bible, and became an essentialized figure in the role of a masculinized demonic Other. This was a deliberate imposition of Eurocentric ideology with the intention of erasing Indigenous Yoruba meanings and philosophy. It is no secret that the explicit goal of both European and native missionaries was to have "uncivilized" Africans know the Christian God and have them jettison their "pagan" ways.

In short, the Yoruba Bible is a classic example of book language that stiffens and overwrites Esu's multifaceted Indigenous role. And this, in essence, is what cosmological imperialism is. Hegemonic notions of Esu as the devil are both historical and contemporary manifestations of Christian fundamentalism and dominant masculinist discourse. Euro-Christian biblical discourse through which many Yoruba elders and youth have encountered Esu continues to shape their ideas concerning Indigenous Yoruba identities. Such perceptions demonstrate the dangerous consequences of colonialist universalisms, and the erroneous hazards of assuming that all worldviews can be measured, judged, and effectively interpreted according to Eurocentric Christian beliefs and values. However, this reality is not to deny members of the Yoruba diasporic community their agency, especially given that while in the minority, not all community members were completely convinced that Esu was the devil. Instead Yoruba elders Sade and Mrs. Fayemi, and youth Bisi, all named the evils of racism and (to a lesser extent) sexism as more harmful and injurious to their being than Esu the Orisa. These interpretations of Esu and the insistence on highlighting racist and gendered oppression are powerful forms of African-centered countercolonial feminist resistance that members of the Yoruba community are engaged in. Especially in the example of Sade and Mrs. Fayemi, this demonstrates their refusal to allow dominant Eurocentric worldview to prescribe how they read the sacred and, ultimately, how they know or understand their Yoruba Indigenous culture and selves.

Moreover, nor is this to deny the agency of those whose perceptions of Esu and Indigenous Yoruba spirituality on a whole were more consistent with dominant Euro-Christian constructions. Although most Yoruba community members featured in this project demonstrated having uncritically accepted Eurocentric worldviews in their perception of Yoruba Indigenous spirituality, there also existed the underlying subtext of dissent because many acknowledged and were astutely aware of racism, and spoke about negotiating the discriminatory conditions of racism and oppression in Canada on a daily basis. Examples mentioned include sexual harassment, struggling for jobs, education, and overall fair social treatment. Hence, the religious compliance of many of the Yoruba elders and youth in this book might be viewed as strategically honing a space for solace, self-preservation, and survival in oppressive and pathologizing contexts. To be clear, I am not saying that Yoruba community members who espoused these problematic views do not genuinely believe Yoruba Indigenous spirituality to be inferior, paganistic, or a Godless religion. Rather, what I am arguing is that they are both compliant *and* dissenting, and that these types of contradictions speak to the nuances of how both uncritical acceptance and resistance to colonial oppression are messy, paradoxical, and ultimately unfinished. Regrettably, Euro-Christian discourses that pathologize Yoruba Indigenous spiritualities dominate in many community members' imaginations. What is often not tapped into or at the least remains concealed and congested is access to the liberating possibilities of decolonization and healing that are embedded in Yoruba Indigenous spirituality. In essence, these are the challenges of embracing one's Indigenous knowledges in Eurocentric Christian contexts: carving out progressive spaces where one can openly and proudly learn about, commune with, and share Yoruba Indigenous culture without the risk of being shamed or ostracized by one's community.

Cosmological Resistance: Recovering the Sacred, Recovering Esu. A Call That Begs the Response

> The devil is not the terror that he is in European folk-lore. He is a powerful trickster who often competes successfully with God. There is a strong suspicion that the devil is an extension of the story-makers while God is the supposedly impregnable white masters, who are nevertheless defeated by the Negroes. (Hurston, 1995: 230)

Placing analyses in cosmological context offers important lessons about the often elusive and messy configurations of resistance, empowerment, and *Ase* (life-force or strong energy). Zora Neale Hurston understood this and appreciated the political and spiritual value of African cosmology as a crucial site from which to explain Black life and experience. I return to this passage from Hurston (1995) because it settles some of the unsettling tensions of blockage in having limited access to one's Indigenous ways of knowing and being. Hurston's insightful rearticulation of the devil highlights her Afrocentric feminist consciousness and serves to mark Eurocentric constructions of Africans and Blackness *as constructions*. She understands the devil in non-Christian terms and in doing so muddies the Euro-Christian dichotomy between good and evil. Similarly, in repositioning Black/African people as tricksters and powerful story-makers that inevitably defeat White masters, Hurston transposes Indigenous African folklore and spirituality as key sites of resistance and empowering spaces that can provide healing to one's being. In effect, her interpretation of the devil is a critical and spiritually literate one that frees Yoruba hesitation around embracing Esu as an Indigenous symbol of colonial subversion and self-determination. And despite the often disheartening challenges of learning and living Yoruba Indigenous knowledges in Euro-dominant contexts, the key imperative is a politics of hope that rests in nurturing the impulse toward wholeness, an impulse that we all have.

Notes

1. See section in this chapter titled "Cosmological Imperialism: Gender and the Essentialized Masculinization of Esu" for more on Orisa's indeterminate gender.

2. For a more in-depth discussion of Esu's roles and responsibilities in the Yoruba cosmos, see the section of this chapter titled "Cosmological Difference: A Struggle Between Two Metaphysical Systems."
3. Although the notion of "the trickster" has often been perceived as a negative description, this is not how I nor Idowu conceptualize the term *with specific reference to Esu*, who, as I have cautioned above, cannot be reduced to simply one role and responsibility in Yoruba cosmology. To do so would be reductionist and conflates this uniquely complex Orisa. It has been suggested by Ulli Beier (2001, 33) and Oyekan Owomoyela (1997, x–xvi) that the trickster role *by itself* is a negative portrayal; however, when keeping in mind Esu's other multiple tasks and responsibilities that operate in concert with the character building agenda behind Esu's "tricks," it is clear that Esu as trickster is in fact *not* a negative description of how this multifaceted Orisa operates in Yoruba cosmology.
4. See Chapter 3 for a more detailed argument of how Yoruba Orisa are reconstituted as idols in Euro-Christian discourse.
5. To give some background to the above quote and Crowther's life, he grew up in the town of Osogun, Oyo, when, in his early teens (approximately 1821), he was kidnapped into slavery. However, a year into his captivity, the British antislavery ship, the *Myrmidon*, attacked and intercepted its Portuguese colleague, the *Esperanza Felix*, and took the 189 enslaved Africans to Sierra Leone. Crowther was subsequently given room and board with missionaries from the Christian Missionary Society (CMS), and as a free man in Sierra Leone was taught to read and write while simultaneously exposed to the word and philosophy of Christianity through the Bible. He was such an avid student that he was able to read the New Testament and appointed a teacher in one of the local schools within six months of his arrival in Sierra Leone (Ajayi, 2001). By 1825, within three years of his new life in Sierra Leone, Ajayi was baptized and had renamed himself after Samuel Crowther (a pioneer of the CMS), officially becoming Samuel Ajayi Crowther.
6. Despite the large body of African/Black feminist work in existence, popular misconceptions of African women not fighting in wars still persist. Mbuya Nehanda, Moremi, Queen Nzingha, Yaa Asantewa, Mary Muthoni Nyanjiru, Oya, the third wife of Sango who was deified to Orisa status upon death, and countless other unsung African women fought alongside African men in wars as both soldiers and military strategists. See Aaronette M. White's "All the Men Are Fighting for Freedom, All the Women Are Mourning Their Men, But Some of Us Carried Guns: Fanon's Psychological Perspectives on War and African Women Combatants" for more on African women in armed struggles. Also, see Oyeronke Oyewumi, *The Invention of Women* and Oyeronke Olajubu, *Women in the Yoruba Religious Sphere* for more detailed analyses and discussion of gender and the contributions of women and/or feminine-based entities, such as the Orisa Osun. Osun is an example of a feminine Yoruba deity that has been excluded and written out of Ifa oral texts and largely Yoruba history as we currently know it. Both Oyewumi (1997) and Olajubu (2003) extensively provide evidence illustrating how these current conceptions of gender in the Yoruba context are largely Western-derived and due to European imperialism.

· 5 ·

"THEY'RE NOT ALWAYS RIGHT BUT THEY'RE OLDER"

The Polemics and Paradoxes of Seniority in Yoruba (Indigenous) Culture

> Could it be that Africa yet awaits discovery? This time, however, in the profound, not the geographical, sense, which makes no sense at all as a claim on any inhabited space. A continent yet waiting to be truly discovered—that is, virtually excavated, all magic and reality, myth and history, warts and beauty marks, as a proposition of universal challenge to facile preconceptions. And the first line of explorers should be—who else?—the indigenes themselves, astonished at what they had always taken for granted, or overlooked … It is important to begin here, since the "discovery" that we urge must be primarily that of self, Africa being obviously in existence first and foremost for Africans before all others
>
> —Wole Soyinka, 2012: 28–29

The above statement was made by Baba Wole Soyinka, Yoruba playwright and poet who in 1986 was awarded the Nobel Prize in Literature. He was the first African conferred this honor and duly so for not only Soyinka, but any African writer artist of such accomplished talent. However, for the purposes of my discussion here, I feature Baba Soyinka as fitting to begin this conversation (and call) for a number of reasons: First, Soyinka understands and utilizes a critical spiritual lens, evidenced in an ardent embrace of his Yoruba Indigeneity, which he proudly shares with the world in his brilliant writings (1959, 1976, 1981, 2012). Second, because he is an elder in the Indigenous African sense where elder hood, or seniority, is equated with respect, much wisdom from experience, pedagogy, guidance, and in many instances, simple awe. For the most part, being an elder is equated with one's chronological age where it is believed that the older one becomes the wiser and more experienced one becomes as well. However, I present a caveat here: Although seniority is principally understood according to chronological age, this is not necessarily always the case. I will later discuss some tensions surrounding seniority and chronological age, but for now my focus is to delve into some of the challenges and paradoxes of seniority as an *explicitly* salient feature of cultural life and daily practice among the Yoruba. Here, I engage this discussion of seniority and elders within the frameworks of African/Black feminist thought, Yoruba worldsense, and African Indigenous spirituality that inform these ways of knowing and being.

This leads me to the third reason I began the conversation here with Baba Soyinka's statement: I am of the impression that Soyinka also is making a call to the souls of Yoruba folk (all Africans really) to begin "learning to remember the things we've learned to forget" (Dillard, 2012). I believe Soyinka is also making a call to dialogue with him concerning his insistence that our redemption as Africans lies in our Indigenous spirituality. It lies in us rediscovering ourselves and our continental Motherland through the powerful matrix of our Indigenous knowledges, cultures, and identities. In other words, Soyinka is insisting that *we* are our own leaders, our own heroines and heroes; *we* are the ones who will rescue ourselves when we are courageous enough to re-turn, re-claim, re-cover, and dis-cover ourselves on our own Indigenous African terms. Soyinka understands that our knowledges are sacred. He understands that the Indigenous faith of the Yoruba is legitimate in its own right and that critical engagement with/in Yoruba worldsense promises empowering possibilities of freedom in (re)discovery of who we are.

As the reader you may wonder exactly how this relates to the discussion of seniority and elders I initially began here. My argument is that seniority and an embracement of one's elders is a salient component of Yoruba worldsense and Indigenous culture (as it is for African culture on a whole). Accordingly the lens of critical spiritual literacy is employed to articulate how seniority figures in Yoruba worldsense and, by extension, underscoring its relevance to the souls of Yoruba folk. This lens centers the necessity for critical dialogue on seniority in terms of how we save and re-member our Indigenous culture and identities. Given the realities of historical and contemporary colonialism and imperialism, I argue that Indigenous notions of seniority and elder hood have been fractured, distorted, and in many ways forgotten in terms of the myriad ways one can forget, that is, through disinterest due to lack of affirming Indigenous pedagogy passed on from one's elders; what I would identify as a lack of inheritance; due to its sibling inheritance of colonially derived narratives that construct African Indigenous spirituality as evil, pagan, idol worshipping, and of the devil; and finally, through spiritual congestion that can sometimes manifest as a closeted spirituality. If one remembers that our elders are our pedagogues charged with the duty of teaching the younger generations about our Indigenous spiritual culture, the significance of seniority and its relationship to spirituality becomes clear.

Likewise, despite formal education's claim to educate our children, African/Black children have some of the largest pushout[1] rates. And those of our children who are able to stick it out in these mainstream Eurocentric educational institutions are often doing so and graduating with little (sometimes zero) knowledge of our African histories, cultures, identities, and spiritualities. This heightens the critical role elders play in African communities. Hence, this is a call, not in a way to reproduce racist blame-the-victim narratives where absolute culpability for our children's high pushout rates is solely cast upon African/Black parents' shoulders (and this is often done at the expense of any critical discussion that contextualize these realities in the fact of systemic discrimination, anti-Black racism, and Eurocentric curriculum). No. I engage this discussion and posit my argument of seniority in the spirit of accountability and responsibility, principles that are central to Indigenous African philosophy and culture. This is where elders take care of, love, nurture, and teach our own in affirming ways that empower our children with equally affirming self-esteem, self-knowledge, and self-love. Hence, some questions remain: What are African/Black children and youth learning about Indigenous spirituality? Who are our elders and what are they currently teaching the youth? What are the possibilities of their pedagogy in the context of dominant colonial frameworks that continue to mask as normal? Portions of the former question have been engaged in earlier chapters of this book. However, here I will revisit this question as it specifically relates to my discussion of elders and seniority as understood by diasporic Yoruba people. For the latter questions, again I use the specific example of the Yoruba in diasporic contexts, primarily because I arrived at them through the voices and experiences of Yoruba elders and youth interviewed for this text. Hence, seniority and the role of elders is theorized by answering the following questions: How is seniority constructed by Yoruba elders and youth? What is the role of seniority within a context that pathologizes Yoruba Indigenous spirituality, and how does this social category figure (i.e., as a help and/or hindrance) in Yoruba teachings and the passing down of traditions?

To summarize, using a critical spiritual literacy lens, I examine the social category of age, or what can loosely be translated as seniority in Indigenous terms. As a salient spiritual and social category in Yoruba culture, I analyze how Yoruba community members understand issues of age, the role of elders and respect for elders, as it culminates in Indigenous Yoruba understandings of seniority. I also merge this with other social relations of power such as gender and class. In doing so, I use Black feminist theory to situate the importance of social location and interlocking oppressions as important analytical tools that are linked to the larger struggle for Indigenous decolonization and resistance. This is significant because the specificity of how social relations of power figure in Indigenous Yoruba terms are then taken up in dominant Eurocentric ones. Included in this discussion is how these categories are both contested and contextual. That is, that they are fluid and can change, shift, and be contradictory based on the narratives of how Yoruba people in the diaspora understand age and seniority.

Understanding Seniority and Gender from a Yoruba Worldsense

In this section, I draw on African/Black feminist literature that prioritizes and engages the intersection of seniority and gender from Indigenous perspectives. Since imperialism and conquest (and the legacies of this) are realities that many Indigenous peoples live with, much Indigenous feminist scholarship has focused on the devastating impact of colonialism; namely, how it has displaced and warped Indigenous constructions of gender. Such arguments provide contextualized explanations of patriarchal dominance and violence against women in Indigenous communities. Oyeronke Oyewumi (1997), Bibi Bakare-Yusuf (2003), and Oyeronke Olajubu (2003) are African/Black feminist scholars whose work addresses the meaning and significance of gender in Indigenous Yoruba contexts. Oyewumi's work addresses the naturalization of Western knowledge production processes in African studies. She specifically focuses on the epistemological underpinnings of Western notions of sex and gender, where gender-specific English is written into gender-free Yoruba as foundational evidence that gender, and particularly the category "woman," does not operate in Yorubaland in the same way it does in Western discourses (x, xi). In Oyewumi's estimation, *seniority*, not gender, is the primary power relation in Yorubaland. Oyewumi's ground-breaking exposition of concealed Western social categories (especially the category of "woman" as universal) is a staunch reminder of the need to remain cognizant of how the naturalization of culturally specific (Western) categories prevails even in discussion and analysis of Indigenous cultures. In this sense, Oyewumi's work acts as a decolonizing model for my argument concerning gender as well as other social categories. Oyewumi's deconstruction of Western universalisms reinstalls the need to use Yoruba Indigenous knowledges as central sites from which to theorize and produce knowledge about Yoruba folk. However, Yoruba feminist Bibi Bakare-Yusuf (2003) contests Oyewumi's position that gender is a Western import and not one of the primary organizing principles in Indigenous Yoruba societies. Bakare-Yusuf argues that Oyewumi reduces Yoruba cultural life and experience to discourse, semiotics, and representation, without taking into account the lived effect of language on embodied subjects as sexuated bodies. For Bakare-Yusuf, Oyewumi fails to address "how agents live through and are positioned within the field of power, language, discourse and social practice" (120). Bakare-Yusuf's attention to power, embodiment, and social experience is key in that not only are they of prime relevance when discussions around sex, gender, and anatomy are raised, they are also important analytical tools necessary for gaining a deeper understanding of what Yoruba Indigenous identities and lived experiences in diasporic contexts entail. Because these features are missing from Oyewumi's study, she is not able to address how the racialized, gendered, and age-identified body interlocks with power. Bakare-Yusuf argues that Oyewumi's failure to seriously consider the social relations of power also means that she will not be able to address the complex nuances of how power figures with respect to seniority and gender. However, Oyewumi's and Bakare-Yusuf's positions are not mutually exclusive; their approaches remind us of intersectionality as a fundamental tool of critical inquiry to tease out how power and privilege shape our experiences and identities. Intersectionality is also critical here because age and seniority are not categories that exist devoid of their entanglements with other social categories, such as gender, race, and class, for example. Intersectionality is also relevant in application of critical spiritual literacy where critical examination of normative assumptions embedded in both dominant and Indigenous cosmologies is employed. For example, Bakare-Yusuf (2003) posits that social relations such as class, gender, status, and age are continually shaped by differences in power, and has the following to say concerning the consequences of such inequities:

> [Oyewumi] cannot discuss the fact that the ideology of seniority is very often used as a way of masking other forms of power relationship. It is in this sense that her theorisation of seniority may be seen as politically dangerous. The vocabulary of seniority often becomes the very form in which sexual abuse and familial (especially for the aya/wife in a lineage) and symbolic violence are couched … where victims are reluctant to challenge the abuser in the name of "disrespecting their senior." (132)

Bakare-Yusuf notes that the Indigenous Yoruba category of seniority can become a *conciliatory* site of patriarchal oppression and disempowerment for the "sexuated" (Bakare-Yusuf, 2003) or "ana-female" (Oyewumi, 1997) body otherwise known as "woman." Yet, according to Oyewumi, there is no Yoruba equivalent to the category of woman in the Western sense of the word and social identity. However, as Bakare-Yusuf points out, what does exist are normative notions of seniority that appear to be inflected with covert assumptions of patriarchy which, in practice, can leave women vulnerable and prone to abuse and violence under the guise of seniority. In other words, critical examination of Indigenous worldsense reveal normative practices demonstrating that while abuse and oppression are not epistemically visible in Yoruba Indigenous philosophy, the marginalization of women is practiced. Therefore, Oyewumi's argument is both right and wrong: right in that gender does not operate in Yorubaland in the same way it does in Western society and discourse, wrong in her denial of gender as a primary social relation of power and inequity in Yoruba society. In the end, Bakare-Yusuf's discussion echoes Dei's (2000) sentiment that "sites of disempowerment" in Indigenous knowledge systems cannot be left unexamined and must be critiqued, albeit with keen cognizance of the larger colonial context. This also underscores the danger(s) of romanticizing our Indigenous systems, otherwise we run the risk of compromising our ability to identify the spaces of inequity that exist for the more vulnerable in our communities such as women, children, those who are differently abled, and sexual minorities.

Perhaps the most useful and flexible discussion of gender in Yoruba contexts is that of Yoruba feminist scholar Oyeronke Olajubu (2003), who understands gender as a dynamic process that is mutably constructed and therefore interdependent on other social systems (7). She states:

> gender as construed by the Yoruba is essentially culture bound and should be differentiated from notions of gender in some other cultures. It is a gender classification

> that is not equivalent to or a consequence of anatomy at all times. Yoruba gender construction is fluid and is modulated by other factors such as seniority (age) and personal achievements (wealth and knowledge acquisition). Its boundaries are constantly shifting, and reconfigurations attend its expressions constantly. (8)

Olajubu's emphasis on the flexibility of Yoruba gender and age-based construction is most relevant because this flexibility allows for discussions of seniority and gender to be taken out of narrow binarized conceptions that either deny or decontextually overemphasize one social category over another as primary in Yoruba culture and society. In this sense, Olajubu's articulation of gender as mutable and process-oriented provides an effective model for how to utilize and understand seniority in Indigenous contexts. However, Olajubu's argument of Yoruba gender construction (and by extension seniority) as culture-bound is somewhat contradictory to her position that it is flexible since culture itself is not static. What Olajubu neglects to mention is how, through the violence of colonialism and imperialism, Eurocentric notions of gender, seniority, and other social categories have influenced how the Yoruba construe these social identities. Further, while Olajubu recognizes that Yoruba gender construction is modulated by other social factors such as age/seniority, wealth, etc., nowhere in her statement does she address colonialism or imperialism. In neglecting to do so, she fails to account for the fact that the fundamental impetus behind colonialism and imperialism was to conquer, destroy, and annihilate Indigenous culture to replace it with that of the European. In this context Indigenous Yoruba constructions of both gender and seniority must have somehow been affected and modulated; that is, in some way, they were shifted and displaced according to patriarchal Eurocentric culture. Regrettably, Olajubu does not engage what I think is a fundamental aspect of the varied flexible manifestations (be they hegemonically Eurocentric, Indigenous, or the complicated entanglements of these) of Yoruba social categories and the processes in which they are constructed. I therefore situate the various social relations that modulate Yoruba Indigeneity as complex and interlocking. I am cognizant of oppressive systems such as colonialism and imperialism that in all likelihood have almost certainly warped Indigenous Yoruba constructions of both seniority and gender.

Ultimately, this complicated debate concerning gender and, by extension, seniority in Yoruba contexts is of significance because it underscores the fact that such a discussion is not solely about gender. Rather, it is also bound up with seniority/age, imperialism, colonialism, embodiment, power, culture, the uncritical acceptance of Western discourse as normative, and how all figure from within Yoruba worldsense. Although Oyewumi (1997) convincingly cautions against the dangers of Western universalisms, Bakare-Yusuf (2003) reminds us that continued focus on differences of privilege and power imbalance are imperative in understanding the embodied and material realities of Yoruba women, men, and society as a whole. Meanwhile, Olajubu's argument that Yoruba gender constructions are flexible, fluid, and process-oriented serves as another significant reminder that Indigenous knowledges are complex and dynamic and cannot be reduced to narrow binaries. Accordingly, the debate on gender, seniority, and how these social categories figure in Yoruba society has not been resolved; and it is not my intention to settle these tensions here. Rather, for me, it is more useful to reflect and build on the works of these scholars by using their positions as analytical tools to expand on the convergent and divergent discussions around seniority and gender as significant social categories in Yoruba society. Accordingly, this allows for serious attention to be paid to the spiritual, material, and embodied realities of how seniority and gender intersect with culture and class within the larger contexts of colonialism and imperialism in the Yoruba worldsense. This is what acts as my guide, or analytical road map if you will, in understanding and explicating how the Yoruba elders and youth featured in this book make sense of these social categories.

"They're not always right but they're older": The Salience of Seniority in Yoruba Life

> There is wisdom and then there is stubbornness. There is faith and then there is procrastination. Because wisdom is there, it doesn't mean that there may not be a blind spot. That's where the younger generation comes in to say, "Hey, you know, I understand what you're saying. Would you look deeper? You don't seem to be helping us." This allows the gap to be bridged. The elder steps into another generation, peeks into its reality, then makes amendments.
>
> —Sobonfu Some, 2003: 48

While academic debates around the seniority and/or gender question remain, what is undisputable is the *explicit* significance of seniority in Yoruba language *and* social life. This is where a code of "respect for elders" is deeply believed and practiced by the Yoruba, as is the norm for many African cultures. This reality was confirmed in interviews with Yoruba elders and youth, all of whom answered that having respect for people older than them was a very important element of their Yoruba culture and upbringing. Mrs. Oladiran was absolute in her position, stating that "[The] Elderly, you need to respect them. Whether you like it or not, you need to respect them. You need to respect them." Mrs. Oladiran's daughters Yinka and Tunmi Oladiran agreed with their mother and both believed respect for elders to be very important. Younger sister Tunmi had the following to say:

> Respect is something that is a given and I think it's like, it's very important. And I think with elders comes wisdom. Like they've lived on this earth for so long, they've been through so many things so when someone that's older tells you something, it's not because—they're telling you because they've been through it. And my mom always, when she's talking she says, "I've been through this and I don't want you to go through that." Um, you don't have to get cut twice for you to know it hurts, so you know what I mean? So when someone gives you advice or tells you something it's not because they're telling you because they don't want you to do whatever it is that you want, but they're telling you that going down that path certain things may happen, so it's just ... well, why go down that path?

For Tunmi, respect for elders is largely connected to her belief in their knowledge and wisdom developed from having more life experience than her. She sees elderly advice as a type of guidance for her to not "go down a path" that one of her elders had previously travailed so that it would spare her the painful experience they had to bear. However, older sister Yinka saw respect for elders as based on the elder's behavior as well as how they treated her. Yinka qualified respect for her elders as reserved for "those who do not disrespect me" and stated, "I refuse to respect somebody who has been rude to me or who I don't like, whether they're older or not. So I don't always practice it."[2] Hence, for Yinka respect is a belief *and action* that she recognizes to be tied to a *practice*, her practice, as well as the elder's behavior. Unlike her mother, Mrs. Oladiran, Yinka practices respect for her elders when that respect is reciprocally accorded her. For Yinka, this code of respect for one's elders is not simply about chronological age. This raises one question: How then is seniority determined? According to eminent Yoruba scholars such as Samuel Johnson (1966), Nathaniel Akinremi Fadipe (1970), and Olupona (1991), seniority and respect for elders *is* largely based on chronological age. Yoruba feminist scholars Oyeronke Oyewumi (1997), Oyeronke Olajubu (2003), and Bibi Bakare-Yusuf (2003) have all noted that in addition to chronological age, seniority is relational. Yet, most Yoruba community members overwhelmingly believed respect for elders to be determined by one's chronological age. However, both Mr. and Mrs. Awoniyi felt that this was not always necessarily the case and that respect could also extend to those who were younger than oneself:

Mrs. Awoniyi: Sometimes some people think it's age-wise, but generally it's not supposed to be age-wise solely because I can respect a little child. Okay, seeing what this child is doing ... okay, this child is doing this at what age!!! So I have respect for that child no matter how old the child is! But some people will see that and say, whatever and dismiss it.

Mr. Awoniyi: I do not see age alone being the term for respect; some people are quite aged and they are the most rotten eggs that you can ever find and yet there are people that are so young, probably in their teens, that really command respect. You do not demand respect. Because you are elderly, you think because you are elderly that you must be respected and you are behaving shabbily? That is demanding it. You should command it.

Mr. and Mrs. Awoniyi recognize that respect for elders is not an absolute that is always determined chronologically, and are able to appreciate the potential of Yoruba youth in having a voice and making important contributions as well. Especially noteworthy is that Mr. and Mrs. Awoniyi offer critiques of this absolutist position, first through highlighting the error *of elders* in dismissing younger members of the Yoruba community, and second through emphasis on the fact that there are elders who do not have good character (what, in Yoruba terms, is known as *Iwa pele)* and are therefore not in a position to impart the knowledge, wisdom, and understanding needed to guide younger generations as it is believed an elder should. In other words, some people who are senior in chronological age do not deserve the title of an elder and the respect that comes with this because they quite simply are "rotten eggs" or engage in shabby and dismissive behavior. While Mr. Awoniyi gives no examples to explain what constitutes "shabby" or "rotten egg" behavior, we are able to understand that he does not believe one's age automatically accords one respect. In his opinion, respect is earned by commanding it through one's behavior and how one treats others. Mrs. Awoniyi also believes chronological age should not be the sole determinant for respect and wisdom. As noted above, she does not appreciate the outright dismissal of youth or young children based simply on their young age because it is then assumed that youth is equated with absolute lack of knowledge: a deficit of sorts. The Awoniyis' position is also reflected in their daughter Bisi, who tells of an experience she had with a Yoruba woman who refused to admit not knowing about a matter that Bisi did have knowledge of. Bisi's experience with an older Yoruba woman below illustrates the error in elders' sometimes dismissive behavior toward youth and younger generations simply based on chronological age:

> I think respect for elders is important but elders overstep their boundary in the sense that they think they can talk down to you. For example, this lady came to my house and she was about 30-something when she came and I was way younger. And we saw a commercial for *Free Willy*, and she's like, "Oh that's a dolphin." And I said, "No, that's a whale" and I walked off. She said, "No, that's a dolphin, I'm an older person and I know this and blah, blah, blah ..." I had to bring the encyclopedia for her and take her to the Internet kind of thing and she stayed arguing with me. I feel like she set precedence for me because I was able to say, look how basic that is. Imagine if it's something else that they're arguing with you about! ... your views don't even count because you're young. And it's a respect factor right? ... So you don't even get any kind of respect from them because you're young and it equals to you are not wise.

Hence, while the theme of respect based solely on chronological age is the dominant determinant for many Yoruba, it is also contested and resisted both by Yoruba youth and elders.

Sociologists N. A. Fadipe (1970: 132) and Oyewumi (1997: 41) have both emphasized the importance of seniority also as a responsibility and not solely a privilege. This was echoed in Yoruba youth's and elders' responses to the question of what role elders play: There was an overwhelming consensus that elders held a number of responsibilities where they were to guide, impart wisdom, teach younger ones, protect, be a role model, lead by example, solve family problems, be a spiritual leader, and tell the truth. However, when I asked what should be done when an elder does something that is wrong, aside from the common response of initial silence, there was a general understanding that challenging, questioning, or speaking to an elder about their wrongdoing needed to be broached carefully. Answers ranged from Sanmi Fayemi, Niyi Olusanmi, and his mother Mrs. Olusanmi answering that they did not know what they would do, to Mrs. Fayemi insisting that we as younger people have no right to "scold" an older person. Mr. Fayemi, Mr. and Mrs. Oladiran, and Sade Oriola all had similar responses: they added that the next elder by chronological age should address the matter and speak to them because "elders make mistakes too." Mrs. Awoniyi advocated speaking to the elder by one's self in private, but "doing it in a respectful way," and Mr. Awoniyi and daughter Bisi both felt there needed to be more accountability and responsibility when elders did something wrong. Meanwhile the other youth, namely Dele, Yinka, and Tunmi, all specified that it depended on what the elder had done because, as Tunmi stated, "They're not always right but they're older."

When I offered the examples of woman abuse or sexual assault both Yinka and Tunmi felt that the elder was no longer an elder in their eyes and they said that they would ignore that person. At first Dele advocated the involvement of older family members and/or the law, but he then recanted by insisting he heard of abuse in other cultures but knew of none in the Yoruba community. When I shared that abuse of these sorts cut across all backgrounds, cultures, races, and ethnicities, including the Yoruba, it seemed Dele began to disengage in the interview and responded with, "mmmhmmm."

Ironically, it is the response of the youth to the question of what to do when an elder does something wrong that was of much interest and required closer examination. For example, Dele's answer to my more specific example of domestic abuse indeed can be read as denial of abuse and a naïve defense of Yoruba elders that are male, a response that undoubtedly warrants deeper analysis. However, Dele's initial response to this question is worthy of closer attention. He states:

> Men who abuse their wives … well I guess it's supposed to be a family kind of thing unless the wife is so unlucky that the elder is the head of the family in general or something. Hopefully there will be somebody else the wife can go to like—or an— … if it can't be resolved within the family then I guess the wife has to leave the husband's family and go back to the original family or maybe even involve the law I guess.

It was after this statement that Dele then felt the need to "defend the men" and claim that "the couple could have been fighting" or that "she could have been kicking his ass." Again, Dele's alignment with patriarchal discourse of denying woman abuse, couching it in hegemonic narratives of "couples fighting" and "men can be abused by their wives/girlfriends too," is clear. However, his response to my question of domestic abuse first began with the correct assumption that it is more likely for women to be abused by their male partners than the reverse *and* that the elder is likely to be male. Dele immediately recognized this and suggested that the woman being abused should attempt to enlist her family's help and *leave* or *involve the law*. Dele immediately thought of the abused woman's safety and knew that (his assumption of) a male elder as the head of the family could likely mean she was not safe, what he called "unlucky." He later recanted and proceeded to spew a "blame the victim" narrative. Yet his first response indicates a level of awareness about the abuse of women *as well as* abuse of power that especially male elders may use to silence younger people (i.e., wives) they are abusing. Dele's brief but powerful answer highlights that he knows more than he claims to when he ended our conversation on this issue and insisted that he did not know of any instances of abuse in the Yoruba community. His first response of focusing on the abused woman's immediate safety, followed with denial and recantation in the form of alignment with patriarchal narratives about abused women, suggests Dele may have been struggling with this question himself, possibly because contrary to his claim, he may have heard of and/or witnessed abuse in the Yoruba community, the type where specifically male elders are maltreating their wives who are almost always younger women. Or Dele could also have been strategically negotiating the urgency of a woman's safety in giving me a pedestrian answer couched in the popular hegemonic discourse of abuse as something to be denied and left in the hands of the couple to deal with because it is a "private" affair. This nevertheless reinforces Bakare-Yusuf's (2003) position that

> the ideology of seniority is very often used as a way of masking other forms of power relationship … [and] often becomes the very form in which sexual abuse and familial (especially for the aya/wife in a lineage) and symbolic violence are couched … where victims are reluctant to challenge the abuser in the name of "disrespecting their senior." … [Hence] seniority often becomes the very institution for the exercise and legitimisation of pernicious forms of abuse. (132)

In other words, seniority as both an ideology and lived social category is much more nuanced and complex than the simple elder/youth dichotomy. It is also imbued with unexamined experiences that, in these particular cases of violence and abuse, are highly gendered. For this reason, responses such as that of Dele suggest that despite seniority as a primary social category in Yoruba daily life and experience, it nevertheless is interlocked with other social relations of power such as gender, wealth, status, etc.

Nuancing Seniority and Gender: The Imperative of Black Feminism and Intersectionality

> It becomes imperative that we bring Oyewumi's account of seniority into dialogue with the experience of sexuated existence … The logic of seniority therefore suggests a fairly strict gender hierarchy in practice, but, one which is, as Oyewumi rightly claims, absent within the Yoruba hermeneutical universe … But if we stay at the level of the explicit meaning and symbolic coding, then we miss out on the gaps, significant silences, and concealed meaning within any particular mode of address. (Bakare-Yusuf, 2003: 128–131)

Here, I extend my prior discussion of concealed meanings to a particular form of silence and gap that emerged during interviews with Yoruba community members. It concerns the often concealed gendered meanings entangled in seniority and the "sexuated existence" (Bakare-Yusuf, 2003) of elders, even though for many Yoruba folks, it may rarely be spoken of as such.

I asked elders and youth if respect for elders is based on gender and all 14 responded that it was not. In fact, many followed their initial response with an explanation that in their experience both male and female elders were accorded respect. For example, via titles such as Baba (father/Daddy), Iya (Mother/Mommy), Aunty, Uncle, Ma (Madame), Sa (Sir),[3] and both male and female elders (determined by chronological age) are greeted in the traditional manner that younger Yoruba were taught to greet those older than them. However, during interviews with younger Yoruba folk Yinka, Tunmi, and Dele, two nuances of seniority as gendered emerged. The first entails a pattern where the "head of the family," which is almost always male, is deemed to be the most respected, most knowledgeable, and therefore the "primary" elder within the family. The second nuance involves the overwhelming presence of only men as the public elders in the Yoruba community, especially when outsiders or non-Yoruba wish to access the community. For example, when I asked Dele who the most respected and most knowledgeable member of his family was, he immediately cited his late grandfather and then said that due to his grandfather's death it was his mother's younger brother. When I asked why, Dele answered that it was because his uncle was the head of the family. I then asked if he was the head because he was male and he laughed saying, "You see the sexism I didn't even notice?" After this exchange, Dele said that his mother also has clout and maybe the head of the family title was shared:

> Well my mom she still has her clout you know what I mean? She's here [in Canada, the West], she sends money back there and yeah, definitely she has her clout, so maybe it's shared actually. It's just an honorary title for my mom's younger brother maybe.

Although Dele's mother Sade indeed has economic clout with their family in both Canada and Nigeria, this was not Dele's first response and it seemed his understanding of the "head of the family" title was embodied by Yoruba men. This was also apparent when I asked Dele if the discussion would be as conflicted if the roles had been reversed between his mother Sade and her younger brother. I asked, who would have been deemed the head of the family then? Dele answered that it would have been "automatic" that his uncle would be the head of the family. Hence, for Dele, not only is the head of the family synonymous with being male, it is also synonymous with seniority and being an elder. Similar to Dele, Yinka answered that respect for elders is not gendered yet openly stated that an older man would likely get more respect than a woman. When I asked why, Yinka responded:

> Because a man is considered to be the head of the family ... so they're the ones who get the respect. Like my mom, she has a um—like she has two older sisters and then an older brother and then an older sister and then her and then a younger brother. So the parents have died when they were young, and I find that her older brother who's the third born is considered the head of the family because he's a man but yet he has siblings that are older than him ... My mom has older sisters and he has older sisters that are maybe 74 and 70 or something and he's maybe in his late 60s but he's still considered the oldest. If there's any problems, my mom and them go to him right. Whenever there's functions like Christmas dinners, New Years, they usually get together and say we'll go to his house. Whenever kids need tuition they go to him right, because he's considered a male figure [and] a male figure is always considered to be the above.

Yinka's discussion here demonstrates a number of silences and gaps that are not explicitly stated in Yoruba language and discourse with respect to age and seniority: First, that there are significant moments when male privilege displaces a gender-neutral form of seniority as determined by chronological age, especially where the familial title of "head of the family" is evoked, used, and practiced. Second, while this "head of the family" title is presumed to be gender-neutral, bound up in this silent presumption is the implicit presence of the male as norm, male as leader, and therefore male as head of the family who is worthy of respect. This then culminates in the male as primary elder and senior, even if he is younger in age than his female siblings. Third is that familial social relations reveal gender to be silently bound up with/in seniority. Embedded within this entanglement is the hegemony of hierarchy where the male Yoruba body and identity is more likely to be considered an elder and therefore deemed more respect worthy than is that of the female Yoruba body. Accordingly, the concealed meanings of gender inequity are present within the silent gaps of the supposed gender absent category and discourse of seniority. In her critique of Oyewumi's misguided insistence that "Yorubas don't do gender" (1997) but instead do seniority, Bakare-Yusuf explained that social dynamics and hegemonic practices often exist below the level of discourse and linguistic pattern as experiences of sexuated existence (128, 130). And clearly from Yinka's and Dele's experiences, this is not an either/or polemic: Yorubas do both, and then some. In other words, seniority interlocks with gender, and other social relations of power such as status, wealth, race, age, etc., in complicated and quite nuanced ways.

The second nuance indicating that seniority is hegemonically gendered comes from Tunmi Oladiran. Although she answered no when I asked if respect for elders is based on gender, she added the observation that "I don't see a lot of elders that are women." When I asked her to elaborate, Tunmi had the following to say:

> Like, if you ever ask who an elder is, they're always referred to as men. Actually the women aren't—Like I've never really, truly never seen a woman elder. Not to say there couldn't be one, but I haven't. Like if you ask for elders in the Nigerian community, the majority of the time you're gonna be directed to a man.

As with Dele and Yinka, Tunmi does not explicitly discuss gender in the term "elder." But it is apparent: lurking, and equated with man. Yoruba man. In this sense, seniority or being an elder becomes a silent male-oriented normative that is loaded with concealed meaning which, once again, are equated with the sexuated existence of the older Yoruba male body as normative and therefore representative for all elder bodies. This is another example that reveals the hierarchy of gendered nuances with/in seniority, and how they are profoundly present yet concealed, absent in the Yoruba language and discourse of *agba* [elder], but nevertheless show up in the social dynamics and hegemonic practices that operate in the daily lives of Yoruba men as privileged. Yoruba women are secondary and therefore unequal; Yoruba youth are socialized to obey and reproduce these gender inflected practices of seniority.

"What Scares Us?": A Polemic Against Fear, Anti-Africanism, and, Yes, Sometimes Against Your Elders

> This weekend I had an intense personal experience with African traditional religion … But unlike many Ghanaians, when it comes to religion, I consider myself a truth seeker so I am open to learning about religions. I tend to think favourably of the Christian faith, perhaps, because I was raised Christian and live in a society that largely favours that religion. But I know that I don't know so I keep myself open to exploring different religions. What is ironic is that even though I claim to be exploring different religions, until this weekend, I wanted nothing to do with

> African traditional religion. Nothing to do with shrines, *mmotia* [something like fairies], *abosom*, and libation, and African spirits … I considered them evil …
>
> … I cannot say for sure that African traditional religion is evil. I cannot say for sure that it is good. I know that I have been preconditioned to consider it evil. I also know that I do not know. I would like to find out, but I'm scared of the whole affair. My fear is an irrational fear. Because every time I confront my fear, I grow. (Cleland, 2009)

The above statement was written by Ghanaian blogger Esi Cleland, who courageously wrote an entry of her first experience with African traditional religion in 2009. While she does not explicitly identify as such, Cleland is challenging Eurocentric constructs of African spirituality that vilify it. Accordingly, she is engaging a central tenet of black feminism: demonstrating her emerging power as an agent of knowledge that fosters the humanity of African peoples as fully human (Collins, 1990). Cleland captures the essence of two key findings I want to focus on here using critical spiritual literacy: *fear* and *discovery*. Fear, I would argue, is the quintessential emotion that has challenged and blocked Cleland, many Yorubas, and Africans in general from discovering their traditional African religions. Evidenced in Christian discourse and doctrine that has threatened a devil-ridden, fire and brimstone afterlife if one does not agree to be saved and commit to Jesus Christ, fear has been a fundamental emotion instrumental to European civilizing projects. Analogous to this has been the endemic fear of African spirituality as evil due to Eurocentric constructions of the like. As I discussed throughout this book, Eurocentric discourses of African spiritually pervade the psyche more often than do any actual material experiences of our Indigenous spirituality. In this sense, fear is a most salient emotion that challenges the learning and acceptance of Indigenous African spirituality. This is demonstrated in the narratives of Mrs. Awoniyi, her daughter Bisi, and Yinka Oladiran, all of whom gave rather stereotypical accounts when asked to share their thoughts and feelings about the Indigenous Yoruba faith:

Mrs. Awoniyi: Eh, ahh, I … ah, that to me … it's a no, no. I don't like it. I can give you an example, ok? Ah, at one time my uncle, they wanted to give him a chieftaincy title. His wife called my mother [to tell her] that her brother is about to take a chieftaincy and my mother just ran there [to his house]. She said, "No, no, no!" The thing is, when you go in, you start eating with them. They have lots of different things, traditional things … if you go into the traditional things, it's not just having a chieftaincy title or being the Oba, there is more to it when you go inside. Whatever they are doing when you go there, that is what you will be doing too.

To them it might be good, but to me … I don't like the way the Obas, the chiefs, the way they do their stuff. I come from a Christian family and we want it to stay this way. Because, at one point they wanted my father to become an Oba and he had to run away. He said he didn't like it and didn't want to be like them.

Bisi Awoniyi: My aunt, when we were in Nigeria, our house was robbed and apparently the people that robbed the house, they went to one of their oogun (Yoruba medicine) men and told him that they should make it so [that] whoever answers the door, [they] just let them in and take their stuff, and that's how they take the stuff. And that's exactly what happened. They came dressed up like they were visitors and cleared out the house. And then my aunt was so affected [by that] until she passed [away]. So when people say it's not real, yes it is because they wouldn't just keep talking about it if it wasn't, and especially because it happened with someone in my family. So that's what I think about Yoruba religion.

Yinka Oladiran: I lived in Nigeria for three years from the age of two to five. So there were different apartments within the house but we had separate houses. And there, there was a lady that lived there, a nice lady, a older lady who um … my mom used to let us go play with her and she used to like me a lot. And my mom used to go to a church at that time, a Pentecostal church that was very spiritual. And there was times when they were doing like a revival at the church and the minister that day kept calling my mom's name, right. He was calling my mom's name like saying, "Yinka's mother" but my mom didn't take it in. But then somebody told her like, "aren't you the one they're talking about?" She's like ah, yeah. Then they're like, ok, why don't you go see this person. So she went to go see this person and when she did, right, [she starts to speak in a loud excited whisper] the person told my mom that somebody she lives near is trying to kill me! So my mom had to fast and pray. She had to fast and pray like I don't know for how long. But what had happened was, it was the lady's child that died. I used to go to her house, we used to live in the same place because everybody knows everybody in the same duplex. Yeah, so it's [traditional Yoruba spirituality is] usually associated with negative things. But I guess there's some people that could use it maybe for money, for monetary gain or something right. But it's not usually used for good things.

I thought it important to revisit the above statements because there is a common thread in these stories: Mrs. Awoniyi, Bisi, and Yinka were not the ones who had immediately experienced the stories they told; rather, it was an elder of some sort (i.e., an aunt, uncle, grandfather, grandmother, parent) who was the protagonist of these stories. Again this is not to deny or debate whether or not these incidents occurred. What concerns me is that in the repeated circulation and re-circulation, in the telling and re-telling of these stories (and the many ones similar to them) they act as powerful and believable accounts that deter youth and younger generations from exploring, re-covering, and discovering their Indigenous spirituality. Why? Because the underlying message is that Indigenous African spirituality is to be feared because it is vile and laden with danger, harm, and possibly even death. Simply put, African spirituality is evil. It is a "no-go" territory, which consequently renders it the status of trepidation and terror. Hence, the caveat is that one is to keep one's distance and stay away for one's own sake.

For especially Yoruba youth, disobedience is strongly admonished because youth are often warned by their elders that, if, by chance, they were to disobey them, they would be opening themselves up to danger and evil, which begets harm, and harm, which begets more fear. These re-circulated constructions pervade the individual and collective psyche. It is a vicious paralyzing cycle. Evil in particular is an entity that both Christian and Muslim doctrine has overtly taught its followers to avoid, and rightly so. However, the question becomes, What gets constructed as that which is "evil" and therefore to be feared? While Christian and Muslim discourses are often found at the center of many wars and battles (often against one another), one thing both of these religious doctrines fundamentally agree on is the supposed inferior and uncouth status of Indigenous African religions. Be it Christianity or Islam, both religions have explicitly constructed African Indigenous spirituality as barbaric, pagan, and essentially the tradition of the uncivilized (Olupona, 1991; Soyinka, 2012). Hence, for many Africans and specifically Yoruba youth, disobedience is out of the question not only because they would be disobeying the doctrine and beliefs of the religious tradition they were raised in (given that most Africans will identify as either Christian or Muslim) but they would also be disobeying the teachings of their elders, who were also likely to have been raised in and embrace the Christian or Muslim faith. Doing so could ultimately be read as disrespecting one's elders. Hence, if as a youth (or as younger than your elders) one were to go against these warnings and teachings, and like Cleland one was to explore their own African Indigenous spirituality, whatever consequences (i.e., harm, evil, or death) came of this disobedience left you who disobeyed the warnings as the sole person to blame. It is precisely this fear of Indigenous Yoruba spirituality that is evoked in Mrs. Awoniyi when she speaks of her dad "running away" from "them," the believers and practitioners of this Indigenous faith. Fear is the underlying emotion also evoked in Bisi when she speaks of her aunt's death as having been caused by armed robbers who were successful because they went to *Oloogun* (a Yoruba medicine man). The implication here is that the planning and execution of the robbery were successful due to the sinister powers of Yoruba spirituality that were relied on by the armed robbers via the Yoruba medicine man. It is this same fear that is stirred up in Yinka when she recounts the attempt made on her life by an older woman who had lost a child similar to Yinka's age at the time. Although I cannot presume to know what this older woman was doing with or to Yinka, again the implication is that "something evil was lurking there." And maybe there was. Ironically, despite the problematic insinuation of "witchcraft" as evil, I do not wrong Yinka's mother for heeding her church's warnings and incessantly praying and fasting as a result because this was done to protect Yinka. However, what concerns me once again is how narratives such as the ones told by Yinka, Mrs. Awoniyi, and Bisi are all woven with the common emotional thread of fear. In this way, such stories become quite dangerous when they are the *only* narratives, or the dominant and most popular accounts of Yoruba spirituality. Told and re-told, circulated and re-circulated within Yoruba circles, these stories leave a significant inscription in the individual and communal psyche in terms of what Indigenous Yoruba spirituality is. The narrative imprint of these stories becomes all the more powerful when one considers the malleability of the young impressionable mind, compounded with the significance of seniority and respect for elders in the Yoruba community. What is one to do when these stories are told as truths, and are the most popular stories one's elders are telling the younger generations? What is one to do when such "truths" are compounded with a strong religious upbringing in the Christian or Muslim faith? Be it suggestions of heinous rituals of "eating with them" that necessitate running away; armed robbers that use the powers of medicine men; or old women (read: witches) seeking to engage in human sacrifice of a child, the common thread in these narratives is malevolence, sin, and crime, all tied together with fear as its emotional, psychic, and spiritual glue. And evil undoubtedly is to be feared. Interesting is that none of the elders or youth actually used the word "fear" in the stories they told. Yet, implicit in them is this message. The fear acts as a necessary caution to stay away and distance oneself from treading the waters of African Indigenous spirituality because these waters are dangerous and can only yield consequence(s) of grave harm to oneself and possibly to others as well.

Yet fear is not value neutral and not above critical inquiry. Upon closer examination of this emotion, fear is also entangled in the messiness of everyday life and the endless social constructs that inform this emotion. In other words, fear has a history and is *produced* within a larger social context that is informed by social relations of power and identity such as class, age, seniority, gender, race, etc. In her examination of fear in the United Kingdom and United States over the past 200 years, historian Joanna Bourke (2006) makes a similar argument:

> Crucially, emotions such as fear do not belong only to individuals or social groups: they mediate between the individual and the social. They are about power relations. Emotions lead to a negotiation of the boundaries between Self and Other or One Community and Another. They align individuals with communities ... The feminist slogan "the personal is political", takes on new meaning in the history of emotional expression. (354)

Bourke also talks about fear as a "civilizing emotion" that forces people to act in safer, "law-abiding and co-operative" ways to gain the approval of others (390). Although she cites anti-smoking and anti–drunk-driving campaigns as examples of the more "civilized behavior" that fear can induce, I extend Bourke's argument to include the civilizing projects of colonial powers, namely the British. Undoubtedly, the definitive goal in the imperial project of British colonials was to vilify Indigenous spirituality and civilize the African native that is constructed as savage, barbaric, and backwards. Such constructs of Indigenous peoples would carry little weight without appealing to fear as a conduit for dissociation from those who were the clear burdened bearers of these racist labels. This dissociation was carried out primarily by elite, Western educated Africans, in order to access the colonial capital that would allow for upward social mobility into the stratified ranks of the "civilized." Understandably, what better persona than that of the native medicine man and/or woman, the Indigenous spiritual believers and practitioners, alongside African (Yoruba) spirituality itself to embody, occupy, and bear the brunt of this fear? Still, not only is this type of fear externally imposed, it is also intergenerational, and has been passed down from generation to generation of Yorubas, often through oral narratives such as the ones recounted by Mrs. Awoniyi, her daughter Bisi, and Yinka Oladiran.

However, there also is a politics of hope I evoke here. It is an evocation culled from the simple fact of colonialism and its civilizing project as undertakings that have never been absolute. They have never been complete.

I speak of the power of agency, resistance, and opening oneself (both as individuals and a collective) up to empowering possibilities that allow for the abandonment, or at minimum, a re-writing of colonial scripts, possibilities that allow us to face our fears so that we can discover ourselves. As an agent of new emerging knowledge, Ghanaian blogger Esi Cleland was aware that she had none of her own experiential knowledge of African traditional spirituality and had the courage to admit this. Empowered with this, she faced her fears and embarked on a journey of self-discovery about her Indigenous faith. Below she asks some crucial questions that allow for a critical unpacking of the fear she nurtured until consciously countering it:

> What exactly is it about African traditional religion that makes us steer clear of it? What scares us? How is it that I know more about Eastern religions than I know about Ghanaian traditional religion? Even from a purely academic standpoint, whatever happened to intellectual curiosity, to open-mindedness? How had I closed off myself completely from understanding such an important pillar of our tradition and culture? Am I ready to find out?
>
> … Thoughts, questions, insights? If you're reading this, I'd like to ask you, what experience, if any do you have with African religion. Should we be exploring these questions, critically examining who we are, or is this a no-go area, better left unexplored? Should I take the next step to visit a shrine? Would you? Why or why not? (Cleland, 2009)

I call attention to Cleland's bravery in having stood in the face of her fears and social opposition, to challenge herself to explore her Indigenous spiritual culture. In the end, Cleland was glad that *she herself* had experienced Ghanaian traditional religion by attending a meeting of traditional believers from the Afrikania Mission[4] at the Accra Cultural Centre. Once there, she reported that the people engaged in traditional Ghanaian dancing, drumming, and prayer with Indigenous musical instruments and songs. Although Cleland was torn between fear and curiosity while attending the meeting, she promised herself to take traditional Ghanaian dance classes, thereby opening herself up to continued discovery of self through an exploration of her Indigenous spirituality and heritage.

Cleland's experience embodies the second finding I spoke of above: discovery. Cleland challenged herself to embark on a discovery of her Indigenous Ghanaian faith. Despite her fears, and despite having been preconditioned to think of it as evil, Cleland was able to recognize that she really *did not know* about her Indigenous religion and set out to explore this for herself. I applaud her bravery, for not only was she audacious in confronting her fears, she was also confronting social disapproval in the form of possibly being ostracized or stigmatized by her Ghanaian peers and elders for daring to tread waters that she openly admitted were popularly understood as evil. Yet Cleland understood that fear is an obstruction that mitigates exploration and discovery. Similar to the type of discovery Wole Soyinka spoke of earlier, this challenge is a call for us as Africans to engage in discovery of ourselves, our people, and our land in all its "magic and reality, myth and history, warts and beauty marks, as a proposition of universal challenge to facile preconceptions" (Soyinka, 2012: 28). It is what Soyinka calls "a discovery of self" where "the indigenes themselves [are] astonished at what they had taken for granted" (28). This is the journey that Cleland began: It could not have occurred had she not confronted her fears and critically challenged them. It also could not have occurred had she not questioned her Christian beliefs and upbringing, no doubt instilled in her by her elders. Like Bisi, Tunmi, and Yinka, elders such as pastors, teachers, parents, aunts, and uncles had likely cautioned Cleland against exploring or learning about her Indigenous African spirituality. Nevertheless, Cleland's journey could not have happened had she in some way not challenged (some or all of) these elders, and turned to *other* African elders of the Indigenous Ghanaian faith. This is not an "either/or" endeavor, but rather it is rooted in the understanding that those of the African Indigenous spiritual faith are *also* elders and members of our African community, and deserve to be acknowledged as such. In so doing, one is opening oneself up to learning from these social and spiritual spaces. This is a form of discovery that is valid, necessary, and waiting …

Our Forgotten Elders: Drawing Wisdom from Yoruba Worldsense

All six of the Yoruba youth featured in *The Souls of Yoruba Folk*, Yinka and Tunmi Oladiran, Seun Fayemi, Niyi Olusanmi, Dele Oriola, and Bisi Awoniyi, had heard of a *Babalawo* and *Iyalawo*. However, all had also reported that they knew none personally and spoke of having no interest in meeting with and/or learning from a *Babalawo* or *Iyalawo*. Sisters Yinka and Tunmi and Bisi cited the *oogun* (medicine) that *Iyalawos* and *Babalawos* were known for as "bad" and mentioned warnings from their parents, aunts, uncles, or grandparents as reason enough to stay away. When I asked the youth if they regarded those who believed in and practiced the Indigenous faith as elders, Yinka,

Tunmi, Seun, and Niyi all had similar answers that can be summarized in Seun's response stating, "maybe in age but nothing else really."[5] However, Bisi and Dele both knew of Wole Soyinka as a famous Yoruba writer and spoke highly of him as a respected elder. When I pointed it out to each of them that Soyinka openly embraced Indigenous Yoruba spirituality, Bisi said she still saw him as a respected elder; meanwhile, interestingly, Dele was of the opinion that it was "not really clear if he [Soyinka] *really* practices it or if it's something he uses for his writing."[6] When I asked Dele to elaborate he simply shrugged his shoulders and said that many Yoruba academics study our traditional culture in their writings but are still avid Christians. However, Soyinka's experience offers telling insight as to why Dele's theory is likely inapplicable to Soyinka, who, in his most recent book, *Of Africa*, demonstrates he is a passionate believer and advocate of his Indigenous Yoruba heritage. Of his childhood life, Soyinka (2012) writes:

> My mother was what we call a petty trader. Next to her shop was a traditional healer, a *babalawo*, whose clinic was the verandah of his mud house, under a lean-to, thus making it quite visible from the frontage of our own shop, where I often sat. My father was a schoolteacher, and it struck me that his, and the babalawo's, operations appeared to share the activity of instruction, so I began to take an illicit interest in his methods. Illicit because, to a well-bought-up child from a Christian home, such activities were clearly the work of the devil. Beyond a neighbourly good morning, there was hardly any social intercourse between the healer and our own corner of the block. There was no social ostracism, as both sides interacted in numerous ways, especially in town affairs and trade matters, but we, the children, were strictly forbidden to stray in his direction or play with the children of that household—they were pagans! (108)

The theme of Indigenous Yoruba heritage as illicitly evil, and therefore a justified necessity in keeping one's children away from its followers (i.e., babalawos), is clearly present in Soyinka's account of his childhood in Abeokuta, Yorubaland. Implicit in this excerpt is the idea of respectability, a position that Soyinka's mother and especially his father righteously held, and staunchly protected, in contrast to the babalawo and his family next door, who this respect was not extended to, at least overtly so.

Nevertheless, what is jarringly clear is Yinka, Tunmi, Dele, Seun, Bisi, and Niyi's sharp disconnect from knowing, understanding, and interacting with *Babalawos*, *Iyalawos*, and any other believers of the Indigenous Yoruba faith *as elders*. Unlike Soyinka, the youth in *The Souls of Yoruba Folk* had zero interaction with Indigenous Yoruba healers and diviners. At least, this is what they communicated to me. Combined with the persistent production of Indigenous spirituality as evil and bad medicine, it is no wonder there was no interest among the Yoruba youth in wanting to explore or discover their Indigenous faith. The narratives these youth offer concerning their perception of elders tells a *collective* story that not only reproduces racist colonial scripts of Indigenous spirituality as malevolent and therefore to be justifiably feared, but it also limits their conception of seniority and elders to a narrow Eurocentric lens. This is a lens that is not anchored in a Yoruba worldsense because, from this standpoint, not only are Iyalawos and Babalawos understood to be elders that are accorded much respect, but so also are the un/seen elders such as one's ancestors who have passed on and transitioned into pure spirit, and also the Orisa, Yoruba deities, some of which were once human and became deified through death, while others manifest as forces of nature and are venerated as sacred. Both ancestors and Orisa are sacred beings, crucial elements of Yoruba worldsense and un/seen members of the larger Yoruba (African) community. Yet these sacred entities are forgotten as elders and not conceptualized as such, thereby relegating Indigenous Yoruba understandings of seniority, who elders are, and who they can be to the largely inaccessible peripheries.

The disinterest of Yoruba youth in their Indigenous spiritual heritage is one of the reasons why, in 2005, the United Nations Educational, Scientific and Cultural Organization (UNESCO) declared Indigenous Yoruba spirituality (also known as Ifa) as one of the world's "Intangible Cultural Heritages." In essence, it has been proclaimed one of the world's Indigenous cultures that needs protection because it is at grave risk of not being passed onto younger generations. The UNESCO website states:

> Under the influence of colonial rule and religious pressures, traditional beliefs and practices were discriminated against. The Ifa priests, most of whom are quite old, have only modest means to maintain the tradition, transmit their complex knowledge and train future practitioners. As a result, the youth and the Yoruba people are losing interest in practising and consulting Ifa divination, which goes hand-in-hand with growing intolerance towards traditional divination systems in general. (UNESCO: http://www.unesco.org/culture/ich/en/RL/00146)

Accordingly, I argue that the disinterest of Yoruba folk in their Indigenous heritage, and especially that of Yoruba youth, is due, in part, to a problematic approach to spiritual literacy that is largely uncritical and in many ways blocked. In other words, many Yoruba youth draw on a narrow Eurocentric worldview in reading and understanding the sacred in their lives, and in doing so are cut off from a Yoruba worldsense where *egungun*, *Orisa*, *Ifa*, Iyalawos, Babalawos, and other vital components of Indigenous ways of knowing and being in the world are jettisoned, yet patiently waiting to be discovered. I therefore underscore the urgency of *critical* spiritual literacy to achieve freedom to commune with and recognize the sacred in our daily lives, and especially in those forgotten, blocked, and closeted undercurrents that are the subtexts of our lives. In my attempt to shift our notions of what spiritual literacy is and does, my contribution is an offering in the form of critical spiritual literacy as an excavating tool that will assist us in "learning to (re)member the things we've learned to forget" (Dillard, 2012).

Hence, while many of the conventional Yoruba elders are outwardly disparaging, yet closeting their interactions and engaging in secret with their Indigenous Yoruba spirituality, some of their children are able to assemble a glimpse of it, while many are not even privy to this closeted experience. The collective theme here is of a displaced Yoruba worldsense that has been recast as inferior to that of the European and/or Muslim religious worldview(s). This has rendered a parochial understanding of seniority that does not consider Yoruba worldsense as central. Such narrowness allows for the erroneous dismissal of Yoruba Indigenous healers such as *Iyalawos* and *Babalawos* to be known *as elders*, as *spiritually literate*, as pedagogues and therefore as valuable members of our community who are deserving of our respect.

In essence, re-discovering ourselves takes courage and a willingness to go against what many of our elders have taught us. And these are the elders that may subscribe to Euro-Christian or Muslim superiority and might have taught or told us that our Indigenous culture and spiritual heritage is not sacred. However, discovery necessitates the courage to venture out on our own and explore. It especially necessitates the courage to seek out other elders; elders that we may have never interacted with before; elders whom we may have never thought of or understood as such before. Because, even if they are older, they're not always right. In short, discovery is a call to courageously stand in the face of colonizing and racist constructs of Ifa that show up as fear and explore our Indigenous African spiritual heritage anyway.

Notes

1. See George Dei's *Reconstructing "Dropout": A Critical Ethnography of the Dynamics of Black Students' Disengagement from School* where Dei argues that Black students are not dropouts, but disengage from formal school and are in fact pushed out of school due to systemic

racism, White supremacy, and anti-Black racism prevalent in educational institutions. This is also an important reason why Black students disengage from school, and the vicious cycle repeats itself, with Black students paying the price.

2. Tunmi Oladiran, interview, 4 March 2007.
3. Although the last titles of "Ma" and "Sa" do not necessarily denote the title of an elder, because they are also used for people who are wealthy by their hired help, whom may or may not be older than them.
4. The Afrikania Mission was found by Osofo Okomfo Dr. Kwabena Damuah, a Ghanaian man who obtained his PhD from Howard University. Damuah was a Catholic priest for many years and then re-claimed his Indigenous Ghanaian religion, which gave him the dual titles of Osofo and Okomfo.
5. Seun Fayemi, interview, 25 March 2007.
6. Dele Oriola, interview, 10 March 2007.

· 6 ·

THE RESPONSE

"Ohun ti o wa leyin offa, o ju oje lo" [What follows six is more than seven]

> Spirituality does not come from religion. It comes from our soul.
>
> —Anthony Douglas Williams, *Inside the Divine Pattern*

> Even though African countries have achieved "independence," their own spirituality is still very much imprisoned by the cultures that colonized them. It's an interesting dilemma to observe, especially in the people who try to run away from Africa and then go to so-called "modern" countries. They suffer a spiritual crisis, because the identity they have rejected comes back to haunt them. There is a lot of grief and sadness, as people fall out of grace with their own tradition and refuse to admit it.
>
> —Sobonfu Some, 2003: 89

In essence, spirituality is a significant force, a life force that shapes and frames Indigenous cultures. On one hand, because spirituality has a complex history of entanglement with/in centuries of colonialism and imperialism, knowledge production, historical displacement, cultural genocide, and power are all issues to critically consider. On the other hand, closer inquiry necessitates that equal consideration be given to how spirituality is also bound with/in a complicated matrix of agency, negotiation with, and resistance to colonial and imperial hegemony.

Spirituality has profoundly figured in my journey of lived practice and experience of Yoruba Indigenous culture. Accordingly, this has been the central focus of discussion with Yoruba elders and youth in this book. The principal motivation for *The Souls of Yoruba Folk* came from a sense of alienation, miseducation, and knowing little about my Yoruba Indigenous culture. I wondered if my experience was a shared one and if so, I wanted to understand why. I suspected that such experiences were not solely mine but was unsure if this was the case. My goal then became to engage in deeper learning and exploration of Yoruba Indigenous culture in a diaspora that is entangled within a dominant Eurocentric context. I aspired to comprehend how others in Yoruba diasporic communities discuss and make meaning of particularly the spiritual and linguistic dimensions of our Indigenous knowledges and identities. What I found during this intense journey can best be summed up by a Yoruba proverb: "*Ohun ti o wa leyin Offa, o ju Oje lo*" (What follows six is more than seven).

In his work on Yoruba art and the concept of *Ase*, Yoruba scholar Rowland Abiodun's (1994) discussion of this proverb suggests that "we must look beyond what is easily observed if we are to understand something" (69). I found that, before embarking on this personal/political/spiritual journey in the cold world of academia, it appeared that many in the Yoruba community were not particularly interested in Indigenous Yoruba spirituality and seemed to have abandoned this element of their culture as having little or no value in their lives. I observed this to be the case repeatedly in different forms and contexts, and the common thread was that the message of explicit disdain for Indigenous spirituality never seemed to wane within the Yoruba community I knew. While in a real way I found this observation to be true and shared by most Yoruba elders and youth in this book, I also found there was more to this story, and if I sincerely wanted to engage in in-depth learning about Yoruba Indigenous culture and knowledges, I needed to look beyond what I was observing to understand what was taking place, and why. And I found this: As I observed and read, I realized that conventional understandings of the sacred did not adequately or ethically describe and include Indigenous ways of knowing the sacred. I therefore knew that somehow, I needed a different way of engaging this inquiry. I needed a way that would allow me to learn to remember the things that I had learned to forget (Dillard, 2012). And I knew that this learning was more of a collective type than not. In other words, observing solely with the eye is limited to the visual and does not allow one to "see" with other senses or a nonvisual lens. I learned that there is more going on than what can be seen. There is more to this story than its visual linearity. This is what led me to the Yoruba worldsense, and the imperative of understanding its nuances and contours as my guiding framework. It is this worldsense that led me to realize the need for developing a critical component to spiritual literacy as necessary in my being able to effectively tell the story of the souls of Yoruba folk.

In the Time of "More Than Seven": On the Paradoxical Messiness of Complicity and Dissent

Central to many Yoruba folks' preference for Christianity, and the staunch declaration of a Christian identity, is the legacy of imperialism and colonization that Indigenous African peoples continue to be burdened with. And the Yoruba are no exception. This legacy has been amply theorized by many anticolonial and African/Black feminist scholars alike. However, what lies at the heart of community members' embrace with Christianity is the issue of cosmology and worldview, or what Oyewumi (1997) accurately refers to as Yoruba worldsense, which privileges a range of senses instead of the Western tendency to over-privilege the visual. Hence, this book highlights the problem of cosmological encounter for colonized peoples where religious imperialism has occurred. My discussions with Yoruba folk brought to light the conflict and unequal entanglements between the Yoruba and Euro-Christian metaphysical systems demonstrated in community members' understandings of both. Based on dialogue with Yoruba elders and youth, my findings concerning Indigenous Yoruba spirituality were often paradoxical, contradictory, and what I realized to be folded into a larger main finding that was also a contradiction in terms. It became clear that community members embraced hegemonic Euro-Christian constructions of Indigenous Yoruba spirituality as evil, paganistic, and idol worshipping. And this occurred primarily through their understandings of the Orisa Esu. Yet, I also found that outward claims of allegiance to Christianity did not completely erase Yoruba folks' engagements with their Indigenous spirituality; I found that when further explored, there was sometimes a covert engagement that largely happened in secret. This led me to theorizing that Yoruba community members' involvement with their Indigenous spirituality is one that is hidden and closeted, what I termed "spiritual closets." Similar contradictions were also noted upon finding that elders and youth overwhelmingly read the sacred to lie within Christianity. Thus, I was once again led back to the significance of cosmology or worldsense as foundational to the politics of knowing, especially with Indigenous cultures that have been marked by a history of colonialism. I therefore anchored community members' approaches to Yoruba Indigenous spirituality with critical spiritual literacy and used this idea as an analytical tool to explore folks' blockages, when this was the case. I theorized this primarily by revisiting and teasing out peoples' understandings of Yoruba Indigenous knowledges through the figure of Esu. I found Esu to be an Orisa with unique significance because this deity was profoundly demonized by most of the Yoruba community members featured in this book. I argued that this is symbolic of how African Indigenous spirituality and Africans on a whole are also demonized and pathologized in imperial and colonial discourses. Hence, Esu is an important deity not only because of this Orisa's key role and function within Yoruba cosmology, but also because Esu is the Orisa spoken of the most by community members. In this sense, then, the Yoruba deity Esu is key as an icon who brings to light the challenges of learning and engaging Yoruba Indigenous knowledges in Eurocentric contexts. Upon deeper reflection as to why this particular Orisa is repeatedly mentioned, it became clear that this was not coincidental because Esu is the only Orisa who appears in the Yoruba Bible by its Indigenous name, and is consequently transposed into Christianity as the devil and quintessential archetype of evil via the Yoruba Bible. Yet, it slowly became clear that folks' readings of Esu using a Eurocentric and Bible-based Christian vocabulary were still a form of critical spiritual literacy. It slowly became evident that Yoruba elders and youth knew exactly what they were doing: They were using the tools they had and were given to navigate and survive in a land that refused to nurture them and their souls. However, what Yoruba community members in this diaspora do not realize are the spiritual implications and political ramifications of reading Esu in a Euro-Christian vocabulary that assigns this figure to demonic status. What they do not comprehend are the intergenerational implications of compounding fracture and congestion where discourses of harm, danger, and violence (that community members allege as endemic to Indigenous African spirituality) are visited upon us only in uncritically thinking of and accepting these ideas of ourselves as true. Said differently, characterizing a deity, a central Orisa who is not the devil, *as the devil* is a part of what it means to be in a context that requires a devil. Anchored in the intersections of White supremacy and patriarchy, it is a colonial context where Indigenous practices become evil and Christian practices become good. And for Yoruba folk, for Black people, for Africans, that does something to you. That does something to your soul. It deeply compromises our self-knowledge and ability to love ourselves wholly, *on our terms*. It corrupts our abilities in learning to remember the things we've learned to forget (Dillard, 2012) and compromises our self-determination. In other words, it distracts us from a grounded focus on our freedom. A freedom that requires it be informed by self-affirmation. In part, this articulates the challenge of diasporic Indigeneity and making ourselves whole as Africans who are made busy navigating a barren settled land where occupation persists. When affirmation, self-determination, and freedom are the preferred and more liberating preoccupation, the question remains: How do we (re)member this in a way that keeps us alive? These are the messy contradictions of living in diaspora.

For *The Souls of Yoruba Folk: Indigeneity, Race, and Critical Spiritual Literacy in the African Diaspora*, paradox and contradiction continue to emerge. Namely, while seniority and respect for elders was confirmed as salient by Yoruba community members in this text, it was also found that this social category was fluid, relational (Oyewumi, 1997), and subject to power imbalances. In other words, age and seniority are dynamic social categories. They are not static and not a given. For example, although it was primarily the Yoruba elders that were adamant about the prime significance of respect for elders in Yoruba culture, the youth were more likely to hold the view that respect for one's elders was contingent on their elders' behavior and how they treated others. It was also found (again primarily by the youth) that one's social power as an elder could be used to abuse others, namely, those with less power such as wives and especially younger women. Hence, the youth were more likely to question whether such behavior deserved the respect the role of elder commands. Importantly, these findings remind us of the caveat to not romanticize Indigenous knowledges because they also may possess spaces of disempowerment (Dei, 2000), namely for women, children, and ethnic and sexual minorities. I was once again led back to the significance of Yoruba worldsense as foundational not only to the politics of knowing but also as highly relevant to our emotional well-being because it became clear that fear and exploration have a politic as well. And this was primarily because it was fear that held many community members back from exploring and learning about their Indigenous spirituality.

Moreover, just as there were elders who were acknowledged as undeserving of respect yet retained their power as such, so also were there elders that were *not* acknowledged because of their spiritual practice and belief in the Yoruba Indigenous faith. These elders were relegated to the margins of society because of the marginal status of Indigenous Yoruba spirituality. It became clear that accounts of who is and who is not considered an elder were often encrypted with colonial code and script. Yet, in contradiction to this, also present were critiques of elders who abused their social power in this role.

I argue that while on the one hand many folks featured in this book are inclined to demonstrate having uncritically embraced Eurocentric worldview in their perception of Yoruba Indigenous knowledges, parallel to this is the underlying subtext of dissent, or speaking of the Indigenous self and one's Indigenous culture in ways that are aligned with resistance to the Eurocentric discourses they internalize. Even as Yoruba community members' understandings of Indigenous knowledges does fall complicit with colonial and Christian discourses, this complicity is negotiated as a way of self-preservation, that is, to keep one's self safe or protected from harm and injury. In this case, the injury or harm can be read as being excluded from accessing the currency and dominant social rewards that the institution of Christianity as an explicit form of religious identification offers, as well as the risk of being ostracized from one's community if one is to openly embrace Yoruba Indigenous spirituality.[1] Nevertheless, embedded in Yoruba folks' understandings and discussion of their Indigenous spirituality is resistance. It is a way to tell or speak of the self that is not complicit with Euro-colonial Christian worldview, but rather aligned with protest against the very same discourses that folk are complicit to. A number of community members demonstrated this earlier in this book, where they offered insightful critiques of the Christian church as a mundane social routine, a type of performance that does not necessarily nurture one's spirit and inner being. They also admitted that church attendance served the function of gaining respect/ability and social approval within one's community. An additional example of resistance to Euro-Christian colonialism would be Sade, Mrs. Fayemi, and Bisi's refusal to name Esu as evil, and instead identifying racism as such. Although elders and youth who held this perspective were in the minority, it does not minimize the significance of their resistance against Euro-Christian notions of evil. Similarly, Mr. and Mrs. Awoniyi, their daughter Bisi, along with Yinka and Tunmi Oladiran all offered a critique of seniority and respect for elders as something that can at times be taken too far, especially when an elder evokes social status as a tool of power to control and/or abuse those who are younger and/or deemed to have less power. In doing so, Yoruba folk are active agents in their lives because they demonstrate using pertinent strategies for survival in Eurocentric dominant contexts, while simultaneously refusing an all-encompassing or complete disavowal of their Yoruba Indigenous spirituality. Accordingly, spirituality is approached as a dynamic process that community members actively engage with. Whereas dominant religion and/or spirituality can be oppressive, exclusive, and hegemonic, these systems are nevertheless complex processes through which people demonstrate their agency and resistance.

Revisiting the Aims and Objectives of *The Souls of Yoruba Folk*

I have taken up three learning objectives in this book. First, because research on Yoruba Indigeneity in diasporic contexts is scant, the initial objective was to contribute to the production of critical social theory about Yoruba Indigenous knowledges. The second objective has been to initiate an engagement of in-depth critical learning of Yoruba lived experience and understandings of Yoruba worldsense in the context of Euro-dominant culture. And the third learning objective has been to open a space that engages in critical dialogue about Yoruba Indigeneity that is accessible and more affirming than hegemonic discourses that have largely constructed African Indigenous knowledge as inferior, backwards, and in many ways demonic.

1. To produce and contribute to critical social theory about Yoruba (African) Indigenous knowledges

Although literature on the African Atlantic diaspora is well-established in the United States, Britain, and the Caribbean, the African diaspora in Canada is often overlooked and rendered invisible. With the exception of Adeyanju (2000), there is no comprehensive research on the Yoruba diaspora in Toronto that speaks to the specific realities of this particular Yoruba community. Detailed analysis of the nuanced dis/continuities and retentions of Indigenous Yoruba spirituality among Yoruba migrants in Toronto, Canada, has not been undertaken prior to *The Souls of Yoruba Folk*. Such an absence has created a silence around how Yoruba Indigenous culture is reconfigured in the context of colonialism, and how it is affected in the face of balancing familial and economic demands. This text contributes to filling this gap on understanding Yoruba communities in diasporic contexts. I illustrate this in the previous section, where I discuss my main finding of Yoruba folk negotiating their Indigenous identities to be complicit to Euro-colonial discourses while at the same time protesting these discourses. Hence, *The Souls of Yoruba*

Folk is unique not only because of the focus on Yoruba speaking peoples in the diaspora but also because of its relevance to other marginalized diasporic communities who may be socially positioned in similar ways.

2. To engage in in-depth learning and discussion of Yoruba lived experiences and understandings of Indigenous worldsense in the larger context of Euro-dominant culture

This book has employed Indigenous, African/Black feminist, and anticolonial frameworks of analysis, and in doing so has made an important contribution to the production of critical knowledge and scholarship. There is minimal inquiry into how migrants negotiate their Indigenous identities in colonized spaces that are not Indigenous to them. Analysis from this perspective provides a politically grounded and comprehensive understanding of the linkages between diaspora and Indigeneity in spaces that are hegemonic and colonizing. These theoretical and conceptual frameworks also challenge the dominance of anthropological research that has dehumanized and positioned Africans and other Indigenous peoples as objects to be studied, or merely as "sources of data" (Rosenberg, 2000; Smith, 1999). By contrast, this book is anchored in and part of a larger counter-colonial decolonizing project that seeks to give voice to those who have been marginalized, silenced, and erased (Amadiume, 1997; Lorde, 1994; Oyewumi, 1997; Smith, 1999).

3. To open up accessible and affirming spaces that engage in critical dialogue about Yoruba Indigenous knowledges

In challenging the silencing of Yoruba Indigenous identities, it is my hope that this book has opened up a space for critical dialogue around the politics of how African Indigenous identities continue to be hegemonically constructed in colonizing contexts. Freire and Macedo's (1995) theorization of dialogue offers an approach that suggests it as an epistemological process:

> In order to understand the meaning of dialogical practice, we have put aside the simplistic understanding of dialogue as a mere technique … dialogue characterizes an epistemological relationship … [it] is a way of knowing and … I engage in dialogue because I recognize the social and not merely the individualistic character of the process of knowing. In this sense, dialogue presents itself as an indispensable component of the process of both learning and knowing. (379)

This book serves as an example of engaging in critical dialogue to illustrate the vital role Indigenous culture and knowledges play in decolonizing projects. Dialogue such as this also opens up a space for imagining how these knowledges can be used as a central politic for progressive change, resistance, and especially as a foundation to nurture the impulse toward wholeness that colonialism disrupts.

At this "final" point of *The Souls of Yoruba Folk: Indigeneity, Race, and Critical Spiritual Literacy in the African Diaspora*, I hope the reader has learned about spirituality and religion as central to Yoruba diasporic life, in terms of their role and function as sites of inequity *and* resistance to that inequity. While it is found that historical colonialisms have set the framework for the continued hegemonic construction of Yoruba Indigenous knowledges and identities, this reality does not preclude the critical spiritual literacy of Yoruba folk in the diaspora who employ c/overt strategies in deconstructing religious dominance. Said differently, the colonizing force of European dominant culture is not absolute, nor is it final. This is evidenced in the agency of the Yoruba folk featured in this book and how they negotiate their Indigeneity in the face of social inequity and colonialism. This text demonstrates the gravity of how Euro-Christian discourse influences the lives of Yoruba people. And what is urgently at stake are the long-standing political and spiritual implications of continued marginalization and social inequity: Understanding that the defamation and transposition of African Indigenous practices as evil and Christian practices as good is fundamentally and politically dangerous because it inflicts intergenerational harm to the souls of Yoruba folk, Black folk, and African folk on a whole. This is where the real harm and injury occurs.

Yet, on the shoulders of those who came before me, I embrace a politic of hope and believe in the promise of freedom. If, as the Yoruba proverb suggests, one looks beyond what is easily observed, we must hold onto and build on the incompleteness of a total pathology and demonization of African Indigenous spirituality. The significance of our Indigenous faith and the resilience of the people who, in contradictory and paradoxical ways, refuse to let our spirituality die is that this story is not over; there is more to follow. Although marginalized, African spirituality "moves at those margins" (Morrison, 2008). It does not die.

I thus return to the Yoruba proverb, "What follows six is more than seven," knowing that there is more to follow. And it is our charge that what follows is work that is empowering and self-determining. There is more work to be done with respect to how African Indigenous spirituality is perceived and constructed in the public imagination. And this includes the many Africans who, strategically or not, have internalized colonialist views of the Indigenous African faith. While messy, thorny, and complicated, this is where the call to radically change the oppressive systems in which we live begins. It rests in knowing that despite being daunted by the often-overwhelming nature of what we are up against, liberation is the only choice, and the ultimate prize. Indeed, there is so much more to follow six than seven.

This is my call to the souls of Yoruba folk.

Note

1. Consider the social and economic privileges borne of the currency that Christian identification offers, for example, tax exemptions and the option of having one's children attend a Christian public school that is provided through government funding. There is also access to scholarships to private schools, often only available through Christian organizations. There is also the social respectability of Christian identification, which provides a sense of belonging, not to mention the power, prestige, and status of being in a position of leadership as a pastor, priest, deacon, deaconess, the wife of a pastor, etc.

· 7 ·

AN OPEN LETTER TO TEACHERS

Pedagogical Implications and Applications of *The Souls of Yoruba Folk*

> In order for us as poor and oppressed people to become a part of society that is meaningful, the system under which we now exist has to be radically changed. This means that we are going to have to learn to think in radical terms. I use the term radical in its original meaning—getting down to and understanding the root cause. It means facing a system that does not lend itself to your needs and devising means by which you change that system. That is easier said than done. But one of the things that has to be faced is, in the process of wanting to change that system, how much have we got to do to find out who we are, where we come from and where we are going?
>
> —Ella Baker as cited in Ransby, 2003

Dear teachers,

I hope this letter finds you well.

I write to you, my fellow educators, because there is a role you play in shaping the souls of Yoruba folk.

I began this letter to you with the above statement from Ella Baker because it is powerful. And it is powerful because it is located in taking oppressive systems to task by insisting that they require radical change. Undoubtedly, our educational institutions are one of these systems that require this change because, simply put, it is failing our children. And it is especially failing Black children.

As the people who will spend as much and sometimes more time with Yoruba youth than their parents, you have a role to play, and are implicated in this call as well. There is a Yoruba proverb that states, "*Oju merin lon bi omo. Igba oju lon to*," meaning, it takes four eyes to conceive a child, but 400 to raise her or him. In other words, children are the wealth of life and highly regarded as indispensably valuable in the context of the African community. The Yoruba/African approach to raising children undoubtedly includes their education. However, it is an education that must be sound, and one that does not make for a splitting up of a child's mind from the body and soul. A Yoruba worldsense requires the seamless inclusion of teachers and educators as part of the community. As teachers, you are a part of those 400 eyes that raise our children and so you have a responsibility to heed this call as well.

You may or may not be Yoruba. You may or may not be African and you may or may not be Black, yet you shape and influence how our children understand, know, and ultimately how they love themselves because, as educators, you are involved in the production of knowledge, how it is constructed, how it is learned, and how it is taught. Please understand that this is not an accusatory or confrontational letter. It is a letter of responsibility and accountability, coming from a place of compassion to highlight that, as teachers, you are called to an important charge: to rise to the challenge of teaching from a critical place where social justice and equity are centered and central to your pedagogical practice. And this means that spirit and acknowledgment of a child's soul in their education is paramount. As part of the 400 eyes that raise Yoruba/African/Black children, responsibility demands it. You are therefore deeply implicated in the project of this book and its aims for equity and social justice.

As a Yoruba woman doing this work, that is, as an educator who is Yoruba, woman, scholar, and most important of all, a mother, I often think of these multiple identities and recognize that the common thread woven through all of them is the responsibility of both teaching and learning. More than ever, with the advent of this book was its culmination with my journey of becoming a mother. In this way, it made protecting and lovingly nurturing the souls of especially young Yoruba folk all the more urgent. I often think of the many young minds that I have contributed to shaping as an educator, and the many more I will continue to influence in this role. I am left wondering what type of contribution or effect that will be. An immediate and stanchly confident answer to this question I do not have. Yet, what I do know is that *The Souls of Yoruba Folk* is pedagogically important for us educators in our teaching and learning because in the many years it has taken me to complete this book, I am astutely aware that it is not simply young minds that we educators and teachers shape. Rather, it is a whole person, a human being whose spirit, heart, and soul is part and parcel of the formal schooling process and larger educational experience. Because the Yoruba community members featured in this book negotiated their Indigenous identities in ways that regulated their Indigeneity as a racialized people and culture, and because critical spiritual literacy was often blocked or congested by Eurocentric worldview, this project has taught me, over and over again, of the dangers of not engaging students as whole beings, *spiritual* beings who have social, not exclusively academic, lives outside the classroom. This book has confirmed for me the importance of making what is taught and learned in the classroom relevant to students' lives and experiences beyond that space. Formal schooling needs to be a place where, as educators, we do not continue to replicate and work from Cartesian models of pedagogy that ideologically and philosophically depoliticize, marginalize, or closet students' physical bodies and the dominant social and political meanings that are ascribed to them. What I am speaking of here is the social and political significance of our racialized, gendered, classed, religious, and spiritually based identities and the hegemonic Eurocentric inequities that shape them. Cartesian informed pedagogies (which have been normalized in too many educational settings) dichotomize the mind and body. In this model, the question of spirit, or the human soul, is given no consideration at all because the model is limited and reduces humans to a parochial mind/body split. bell hooks (1994), writing in *Teaching to Transgress: Education as the Practice of Freedom*, discusses the Cartesian informed Western academy and comes to her own realization that she needed to "make a distinction between the practice of being an intellectual/teacher and one's role as a member of the academic profession":

> It was difficult to maintain fidelity to the idea of the intellectual as someone who sought to be whole—well-grounded in a context where there was little emphasis on spiritual well-being, on care of the soul. Indeed, the objectification of the teacher within bourgeois educational structures seemed to denigrate notions of wholeness and uphold the idea of a mind/body split, one that promotes and supports compartmentalization (16).

In contrast to narrow Cartesian models of pedagogy, I ask educators to consider spirit and spirituality as fundamental to one's teaching practice. In particular, *The Souls of Yoruba Folk* offers the position that Indigenous spirituality is a powerful anticolonial knowledge source from which all students can learn different ways of understanding the world and living their spirituality. This includes engaging with their environment and the larger world in different and more harmonious ways. I qualify this with the caveat that in no way am I encouraging the unethical mining and cultural appropriation of Indigenous culture and spirituality in the way it is often taken up in especially New Age discourses and practice. What I am advocating and offering is an opportunity for educators to critically engage the politics of religion and spirituality and to consider how they are informed by social and historical inequities and imbalances of power. In more specific terms, I am asking educators, teachers, and intellectuals alike to reconsider the politics of Indigenous African spirituality and how dominant understandings of this framework are largely shaped by Euro-Christian hegemony. The school classroom can be a powerful space or site for students to begin decolonizing and learn Indigenous-based understandings and approaches to spirituality, in the stead of what is most often drawn from where popular ideas and perceptions of Indigenous peoples are concerned: dominant Eurocentric ideology and worldview.

The first implication of this book is that a shift toward a critical pedagogy of justice is required. What I mean is that *The Souls of Yoruba Folk* offers an example of what teaching from a social justice framework looks like and can be. And this is done primarily through critical spiritual literacy as a lens that centers Indigenous worldsense, equity, and the sacred by examining how the sacred is embedded in our daily lives and experiences, and especially that of our students. The second implication, and my request of educators and teachers alike, would be that a paradigm shift is required where formal education's approaches to community and pedagogy are concerned. This shift is two-fold and inter-reliant. First, the shift must be philosophical, which is informed by and followed with the practical. I say this because Indigenous African spirituality is rooted in Indigenous philosophy and cosmology where interconnection and interdependence are foremost. With respect to pedagogy, this therefore means that there is no separation of the teacher from the learner; the two are connected because they belong to and share the same community, so much so that the failure of the student is the failure of the teacher. Accordingly, the learner cannot fail because there is a connection that is implicit in Indigenous understandings of both the teacher and learner. This connection then becomes the driving force behind the teacher's commitment to a learner's success because their student's success is also theirs. However, this Indigenous African understanding of the connection between teacher and learner is currently lacking where formal education and schooling is concerned, hence the urgent need for a paradigm shift. It is a type of shift that necessitates revisiting and disrupting dominant Eurocentric constructions of community that are parochial, and extending them beyond convention to include the teacher in formal schools as part of the larger community schooling exists within. In this sense, the teacher then becomes responsible and accountable for the learner's success, as is the case in Indigenous African contexts.

To extend my discussion of seniority and elders to this present dialogue, in African Indigenous contexts the pedagogue or teacher is often the living Elder. These Elders are the ones in the community who carry vital knowledge that they are charged with passing on to the following generations. In dominant and conventional contexts, however, Indigenous Elders are not formally trained and consequently not accepted as real teachers. Yet, a departure from teachers in the Indigenous context carries spiritually injurious consequences as well as other dangers, particularly for African students/learners who are highly marginalized and discriminated against in formal schooling contexts due to systemic anti-Black racism and oppression. For them, there is a disconnect and stratified dichotomous split between the learner's academic self (mind) and the whole self (spirit, body, emotions, and soul). I therefore urge you, my fellow educators, to understand the necessity for a philosophical paradigm shift in terms of who is deemed to be a teacher, or what the notion of a teacher is. A pedagogue needs to be understood beyond the narrow confines of only the licensed teacher, or university/college professor, and made to include Elders: the teachers and keepers of knowledge in Indigenous communal contexts. This inclusive approach does not take away from the professor or formally licensed teacher's expertise; rather, it expands on and enriches it in such a way that both the knowledge base and pedagogical practice offered and taught to students is now more equitable, diverse, and relevant to students' lives. And this holds the promise of improving a learner's probability for success. This also has an inspiring impact on both the student and teacher. To reiterate, the student's success is the teacher's *and* Elder's.

The overall implication here is that the paradigm shift must be one that begins with Indigenous worldsense, again, because it is cosmology that provides the philosophical framework for understanding the world around us, and how the sacred is embedded within it. Especially significant is knowing how that framework is anchored in a particular understanding of community because from an Indigenous perspective one's community includes ancestors, living human beings, and those yet to be born. Therefore, this approach carries some important pedagogical lessons. For example, in her article "Rootedness: The

Ancestor as Foundation," Toni Morrison (1984) discusses the presence of an ancestor in African American writing as "timeless people whose relationships to the characters are benevolent, instructive, and protective, and provide a certain kind of wisdom" (343). Inversely, she also discusses the consequences of the absence of ancestors as destructive and a key determinant to the success or happiness of both the character and the writing or story that frames the character(s). Morrison argues that killing one's ancestors is equivalent to killing oneself and being lost because the "conscious historical connection is lost" (344). I would argue that this is also the case in the context of pedagogy and formal schooling, where a learner's success is linked with/in one's community and therefore one's ancestors who provide that "conscious historical connection" to one's self. This notion of self is foundational to understanding a student's current living self because, as discussed above, ancestors are part of one's community within African cosmology. Hence, to excise ancestors indexes another type of congestion or blocked critical spiritual literacy. Said differently, ancestors, the unborn, and the living fleshy human self are all important threads that come together to make a circle whole where, in essence, there is no beginning or end, but rather a powerful continuity of life through transmutation. In this sense, then, the shape of a circle reflects how the individual exists within the context of the larger community and how the living self is intertwined with the worlds of the ancestors and the unborn. The circle is both literally and figuratively a symbol of eternity in its continuity because the self is conceived of as a layering of many synthesized selves that cross over, move in, move out, and move in between the spiritual and material energy fields. Ultimately, the self is an existence extended throughout the cosmos, as opposed to simply being restricted to the material world of human beings.

Consequently, pedagogy that only focuses on the learner's academic self does so to the detriment of the body and soul, and dangerously reproduces the damaging Cartesian model of learning where students are forced to leave their soul and political body outside the classroom door. This type of pedagogy endorses social amnesia and a blocked spiritual literacy where one's own history is forgotten and deemed unimportant. From an African Indigenous context you cannot simply teach a learner's mind without engaging the body. And you cannot teach the learner's body and mind without engaging the soul. You cannot engage the mind, body, or spirit without the ancestor or ancestral self, which represents a conscious engagement with and knowledge of one's history. And these particular philosophical approaches to one's self and larger community are foundational tenets in African spirituality, as anchored in African Indigenous worldsense. To teach these elements without the larger framework of African spirituality and cosmology is to decontextualize and fragment the knowledge, which is interwoven and interconnected. Hence, African Indigenous spirituality and cosmology are cornerstones to African students' success and understanding of their whole self and identity.

This raises the important question concerning *how* one teaches Indigenous African spirituality and cosmology to Black/African students who most likely will identify as either Christian or Muslim. Although the impetus and learning objective of *The Souls of Yoruba Folk* is to engage critical anticolonial dialogue about Yoruba Indigenous spirituality and knowledges, it is not my intention to dichotomize Yoruba (African) spirituality as "good" and Christianity or Islam as "bad." It is possible to teach and discuss African Indigenous spirituality without framing it in a hierarchical either/or binary, which, ironically enough, both the Christian and Muslim colonizing traditions have done with this Indigenous faith. Again, my main finding in this book was that Yoruba community members' resistance to, and complicity with, colonialist discourses is contradictory, messy, and full of paradox. Such strategies employed by these folk demonstrate that a simple either/or paradigm is insufficient in the pedagogical context of teaching and learning about African Indigenous knowledges and spirituality. Much like the Yoruba elders and youth in this text, students' lives will also carry contradictions, paradox, ambiguities, and complexities that cannot be reduced to a simple either/or paradigm, even though students may identify with dominant colonial religions. Formal education and pedagogy, and the classrooms they occur in, offer the empowering possibility of acting as critical spaces in which students can decolonize their minds, mend their souls, and be taught to deconstruct the many racist myths and stereotypes about African spirituality that pervade the public imagination. In this way, counterhegemonic and anticolonial perspectives about African spirituality that have largely been rendered invisible become accessible, are validated, and move to the center. This is what I offer in the form of this book: a critical sense-based approach to education where African Indigenous knowledges and spirituality are being taught from an affirming perspective.

As a mother, Yoruba woman, and educator doing this work, I am excited about contributing to the reclamation of Indigenous (African) Yoruba epistemologies as valid. More specifically, my contribution rests in challenging academic circles to open up to a growing body of knowledge about Indigenous peoples and decolonizing methodologies. Hence, for you, my fellow educators, this work offers a number of additional pedagogical implications in terms of how you can apply this text in your teaching practice. *The Souls of Yoruba Folk: Indigeneity, Race, and Critical Spiritual Literacy in the African Diaspora*

- Offers new articulations and understandings of Yoruba (African) Indigeneity in diasporic and dominant Eurocentric contexts
- Theorizes how Yoruba peoples in the diaspora negotiate their Indigenous culture and identities in complicit yet resistant ways against dominant Eurocentric discourses
- Presents new readings of spiritual literacy, where critical spiritual literacy that makes Indigenous worldsense, equity, and the sacred as central is proposed as crucial to knowledge production, teaching, and learning
- Contributes to opening a space where critical and equity-centered teaching and learning about Yoruba (African) Indigenous spirituality and knowledges are accessible and affirmed
- Suggests the application of key elements of Indigenous philosophy to mend disconnection
- Fosters links between individuals and the larger teaching and learning community
- Suggests collaboration between researchers, educators, elders, and other members of Indigenous communities to increase understanding, avoid cultural appropriation, and avoid the reproduction of oppressive colonial discourse
- Can be used in faculties of education/teacher's colleges for topics on spirituality in contexts of schooling, education, teaching, and research, for example, in courses such as School and Society and Critical and Multicultural Education
- Is beneficial to educators in the areas of sociology, Canadian studies, as well as antiracism, gender, and equity studies because it addresses the need for investigation into the material and spiritual consequences of racism and other forms of oppression that affect racially marginalized peoples and their Indigenous knowledges

There is the need for opening up spaces where critical anti-oppressive, anticolonial dialogue and learning can occur, so that the enduring imperial legacies concerning Indigenous spiritual knowledges are deconstructed and eradicated. The margins in the education system that have relegated Yoruba/African/Black children to the bottom of its oppressive hierarchy must be erased. A powerful entry point lies in developing more empowering experiences and discoveries of African spirituality for African children that are affirming and their own, rather than recycled hegemonic constructions borne from Eurocentric culture. Education that is grounded in Indigenous African-centered feminist and anticolonial frameworks will create more opportunities for this to occur. This is where the call to radically change the oppressive education system that our children are forced to learn in begins. It rests in a politics of hope, a pedagogy of justice, and knowing that it is up to us to devise the means of changing this system.

In the form of our work, teachers are undoubtedly a part of the 400 eyes that raise children of African heritage in the classrooms of our schools. And we need to do better. Our responsibility demands it. A socially just work ethic demands it. And, most important of all, the souls of our children as student folk demand it.

BIBLIOGRAPHY

Abimbola, W. (1976). *Ifa: An exposition of Ifa literary corpus*. Ibadan: Oxford University Press.

Abimbola, W. (1977). *Ifá divination poetry*. New York: NOK Publishers.

Abimbola, W. (1997a). *Ifa will mend our broken world: Thoughts on Yoruba religion and culture in Africa and the diaspora*. Roxbury: AIM Books.

Abimbola, W. (1997b). Language and gender. In *Ifa will mend our broken world* (pp. 143–161). Roxbury: Aim Books.

Abiodun, R. (1994, July). Understanding Yoruba art and aesthetics: The concept of Ase. *African Arts*, *27*(3), 68–103.

Adewale-Somadhi, A. (2001). *Fama's Ede Awo (Orisa Yoruba dictionary, Revised and Expanded edition)*. San Bernardino: Ile Orunmila Communications.

Adeyanju, C. T. (2000). *Transnational social fields of the Yoruba in Toronto*. Unpublished master's thesis. University of Guelph.

Ajayi, J. F. A. (2001). *A patriot to the core: Bishop Ajayi Crowther*. Ibadan: Spectrum Books.

Alexander, M. J. (2005). *Pedagogies of crossing: Meditations on feminism, sexual politics, memory and the sacred*. London: Duke University Press.

Amadiume, I. (1987). *Male daughters, female husbands: Gender and sex in an African society*. London: Zed Books Ltd.

Amadiume, I. (1997). *Reinventing Africa: Matriarchy, religion and culture*. London: Zed Books Ltd.

Amadiume, I. (2000). *Daughters of the goddess, daughters of imperialism: African women, culture, power and democracy*. New York: Zed Books.

Asante, M. K. (2003). *Afrocentrism: The theory of social change*. Chicago: African-American Images.

Asante, M. K., & Mazama, A. (2009). *Encyclopedia of African religion*. Thousand Oaks: Sage Publications.

Awolalu, J. O. (1979). *Yoruba beliefs and sacrificial rites*. Essex: Longman Group.

Bakare-Yusuf, B. (2003). "Yorubas don't do gender": A critical review of Oyeronke Oyewumi's *The Invention of Women: Making an African Sense of Western Gender Discourses*. *African Identities*, *1*(1), 119–140.

Bakare-Yusuf, B., & Weate, J. (2005). Ojuelegba: The sacred profanities of a West African crossroad. In T. Falola & S. Salm (Eds.), *Urbanization and African cultures* (pp. 323–340). Durham: North Carolina Academic Press.

Bascom, W. (1969). *Ifa divination: Communication between gods and men in West Africa*. Bloomington: Indiana University Press.

Bascom, W., & Herskovits, M. J. (1959). *Continuity and change in African Cultures*. Chicago: University of Chicago Press.

Battiste, M. (2002). *Indigenous knowledges and pedagogy in First Nations education: A literature review with recommendations*. Ottawa: Apamuwek.

Battiste, M., & Henderson, J. S. Y. (2000). *Protecting Indigenous knowledge and heritage: A global challenge*. Saskatchewan: Purich Publishing.

Beier, U. (2001). Esu-Elegbara: Ambivalence in Yoruba philosophy. In W. Ogundele (Ed.), *The hunter thinks the monkey is not wise … the monkey is wise, but he has his own logic; a selection of essays* (pp. 29–35). Bayreuth: Bayreuth University.

The Bible Society of Nigeria. (1960). *Bibeli mimo tabi Majemu Lailai ati Titun (The Holy Bible in Yoruba: Old and New Testament) New James Version* (2nd ed.). Lagos: Bible Society of Nigeria.

Bobo, J. (1995). *Black women as cultural readers*. New York: Columbia University Press.

Bourke, J. (2006). *Fear. A cultural history*. Emeryville: Shoemaker & Hoard.

Boyce-Davies, C. (1994). *Black women, writing and identity: Migrations of the subject*. London: Routledge.

Boyce-Davies, C. & Ogundipe-Leslie, M. (1995). Moving beyond boundaries. New York: New York University Press.

Brathwaite, E. K. (1984). *History of the voice: The development of nation language in Anglophone Caribbean poetry*. London: New Beacon Books.

Braziel, J. E. & Mannur, A. (2003). *Theorizing diaspora: A reader*. Malden: Blackwell Publishing.

Brussat, F., & Brussat, M. A. (1996). *Spiritual literacy: Reading the sacred in everyday life*. New York: Touchstone.

Butler, J. (1990). *Gender trouble: Feminism and the subversion of identity*. New York: Routledge.

Carty, L. (1996). Seeing through the eye of difference: A reflection on three research journeys. In H. Gottfried (Ed.), *Feminism and social change: Bridging theory and practice*. Chicago: University of Illinois Press.

Churchill, W. (2003a). *Acts of rebellion: The Ward Churchill reader*. New York: Routledge.

Churchill, W. (2003b). Indigenous strategies of resistance. Keynote speech. York University: Toronto.

Cleland, E. (2009). Confront your fears, discover traditional Ghanaian religion. Retrieved from http://maameous.blogspot.ca/2009/08/confront-your-fearsdiscover-traditional.html

Clifford, J. (1988). *The predicament of culture*. Cambridge: Harvard University Press.

Collins, P. H. (1990). *Black feminist thought: Knowledge, consciousness and the politics of empowerment*. New York: Routledge.

Collins, P. H. (1998). *Fighting words: Black women and the search for justice*. Minneapolis: University of Minnesota Press.

Combahee River Collective. (1983). The Combahee River Collective statement. In B. Smith (Ed.), *Homegirls: A Black feminist anthology* (pp. 272–282). New York: Women of Color Press.

Crenshaw, K. W. (1991). Mapping the margins: Identity politics, intersectionality and violence against women of colour. *Stanford Law Review*, *43*(6), 1241–1299.

Dangarembga, T. (1988). *Nervous conditions*. Seattle: Seal Press.

Dash, J. (1991). *Daughters of the dust* (Film). New York: Geechee Girls Productions.

Davis, A. (1981). *Women, race and class*. New York: Vintage Books.

Dei, G. J. S. (1996). *Anti-racism and education: Theory and practice*. Halifax: Fernwood Publishing.

Dei, G. J. S. (2000). Rethinking the role of Indigenous knowledges in the academy. *International Journal of Inclusive Education*, *4*(2), 111–132.

Dei, G. J. S., & Asgharzadeh, A. (2001). The power of social theory: Towards an anti-colonial discursive framework. *Journal of Educational Thought*, *35*(3), 297–323.

Dei, G. J. S., et al. (1997). *Reconstructing 'drop-out': A critical ethnography of the dynamics of Black students' disengagement from school*. Toronto: University of Toronto Press.

Dei, G. J. S., et al. (Eds.). (2000). *Indigenous knowledges in global contexts: Multiple readings of our world*. Toronto: University of Toronto Press.

Dillard, C. B. (2006). *On spiritual strivings: Transforming an African-American woman's academic life*. Albany: State University of New York Press.

Dillard, C. B. (2012). *Learning to (Re)member the things we've learned to forget: Endarkened feminisms, spirituality, & the sacred nature of research and teaching*. New York: Peter Lang.

Du Bois, W. E. B. (1961). *The souls of Black folk: Essays and sketches*. New York: Fawcett Publications.

Emanuel, A. (2000). *Odun Ifa: Ifa Festival*. Lagos: West African Book Publishers.

Fadipe, N. A. (1970). *The sociology of the Yoruba*. Ibadan: Ibadan University Press.

Falola, T. (1999). *Yoruba gurus: Indigenous production of knowledge in Africa*. Trenton, NJ: Africa World Press.

Fanon, F. (1963). *The wretched of the earth*. New York: Grove Press.

Fanon, F. (1967). *Black skin, white masks*. Trans. C. L. Markmann. New York: Grove Press.

Ferguson, R., et al. (Eds.). (1990). *Out there: Marginalization and contemporary cultures*. New York: New Museum of Contemporary Art.

Freire, P. (2000). *Pedagogy of the oppressed, 30th anniversary edition*. New York: Continuum.

Freire, P., & Macedo, D. (1995). A dialogue: Culture, language, and race. *Harvard Educational Review*, *65*(3), 377–402.

Greene, S. E. (2002). *Sacred sites and the colonial encounter: A history of meaning and memory in Ghana*. Bloomington: Indiana University Press.

Hall, S. (1997). Subjects in history: Making diasporic identities. In *The house that race built* (pp. 289–299). New York: Vintage.
Hall, S. (2003). Cultural identity and diaspora. In J. E. Brazier & A. Mannur (Eds.), *Theorizing diaspora: A reader*. Boston: Blackwell Publishers.
Henry, F., et al. (1998). *The colour of democracy: Racism in Canadian society* (2nd ed.). Scarborough: Thomson Nelson.
Herskovits, M. J. (1941). *The myth of the Negro past*. New York: Harper & Brothers.
Heyd, T. (1995). Indigenous knowledge, emancipation and alienation. *Knowledge and Policy: The International Journal of Knowledge Transfer and Utilization*, 8(1), 63–73.
hooks, b. (1981). *Ain't I a woman: Black women and feminism*. Boston: South End Press.
hooks, b. (1984). *Feminist theory from margin to center*. Boston: South End Press.
hooks, b. (1989). *Talking back: Thinking feminist, thinking Black*. Toronto: Between the Lines.
hooks, b. (1992). *Black looks: Race and representation*. Toronto: Between the Lines.
hooks, b. (1994). *Teaching to transgress: Education as the practice of freedom*. New York: Routledge.
hooks, b. (1995). *Killing rage: Ending racism*. New York: Holt.
Hull, G. T., et al. (1982). *All the women are White, all the Blacks are men, but some of us are brave: Black women's studies*. New York: Feminist Press.
Hurston, Z. N. (1939). *Voodoo gods: An inquiry into native myths and magic in Jamaica and Haiti*. London: J. M. Dent.
Hurston, Z. N. (1978). *Mules and men*. Bloomington: Indiana University Press.
Hurston, Z. N. (1990). *Their eyes were watching God*. New York: Harper & Row.
Hurston, Z. N. (1995). *Hurston: Folklore, memories and other writings: Mules and men, Tell my horse, Dust tracks on a road, selected articles*. New York: Literary Classics of the United States.
Idowu, B. (1996). *Olodumare: God in Yoruba belief*. Nigeria: Longman.
James, J. (1993a). African philosophy, theory and "living thinkers." In J. James & R. Farmer (Eds.), *Spirit, space and survival: African-American women in White academe* (pp. 31–46). New York: Routledge.
James, J. (1993b). Teaching theory, talking community. In J. James & R. Farmer (Eds.), *Spirit, space and survival: African-American women in White academe* (pp. 118–135). New York: Routledge.
James, J. (2000). "Discredited knowledge" in the nonfiction of Toni Morrison. In N. Zack (Ed.), *Women of color and philosophy*. Boston: Blackwell.
James, S. M. (1993). Mothering: A possible Black feminist link to social transformation? In *Theorizing Black feminisms: The visionary pragmatism of Black women*. London: Routledge.
Jenkins, P. (2002). The Next Christianity. *Atlantic Monthly*, 290(3), 53–74.
John, C. (2003). *Clear word and third sight: Folk groundings and diasporic consciousness in African-Caribbean writing*. Durham: Duke University Press.
Johnson, E. P., & Henderson, M. G. (2005). In E. P. Johnson & M. G. Henderson (Eds.), *Black queer studies: A critical anthology*. Durham: Duke University Press.
Johnson, S. (1966). *The history of the Yorubas: From the earliest times to the beginning of the British protectorate*. London: Routledge.
Jones-Jackson, P. (1987). *When roots die: Endangered traditions on the Sea Islands*. Athens: University of Georgia Press.

Kraft, M. (1995). *The African continuum and contemporary African-American writers: Their literary presence and ancestral past*. New York: Peter Lang.

Laguerre, M. S. (1998). *Diasporic citizenship: Haitian Americans in transnational America*. New York: St. Martin's Press.

Last, M. (1981). The importance of knowing about not knowing. *Social Science and Medicine, 15*(B), 387–392.

Lattas, A. (1993). Essentialism, memory and resistance: Aboriginality and the politics of authenticity. *Oceania, 62*, 249–263.

Lawuyi, O. B. (1986). Reality and meaning: A review of the Yoruba concept of Esu. *Afrika and Ubersee, 69*, 299–311.

Lewis, E. (1995). To turn as on a pivot: Writing African Americans into a history of overlapping diasporas. *The American Historical Review, 100*(3), 765–787.

Lorde, A. (1982). *Zami: A new spelling of my name*. San Francisco: Crossing Press.

Lorde, A. (1984). *Sister outsider: Essays and speeches*. San Francisco: Crossing Press.

Mama, A. (1995). *Beyond the masks: Race, gender, subjectivity*. New York: Routledge.

Mama, A. (1997). Sheroes and villains: Conceptualizing colonial and contemporary violence against women in Africa. In J. Alexander & C. T. Mohanty (Eds.), *Feminist genealogies, colonial legacies, democratic futures*. New York: Routledge.

Mbiti, J. S. (1975). *Introduction to African religion* (2nd ed.). Oxford: Heinemann.

McFadden, P. (1998). *Gender in southern Africa: A gendered perspective*. Harare: SAPES Books.

Mikell, G. (1997). Conclusions: African women and state crisis. In G. Mikell (Ed.), *African feminism: The politics of survival in sub-Saharan Africa* (pp. 333–346). Philadelphia: University of Pennsylvania Press.

Morrison, T. (1984). Rootedness: The ancestor as foundation. In M. Evans (Ed.), *Black women writers: 1950–1980, A critical evaluation*. New York: Doubleday.

Morrison, T. (1987). *Beloved*. New York: Plume.

Morrison, T. (2000). Unspeakable things unspoken: The Afro-American presence in American literature. In J. James & T. D. Sharpley-Whiting (Eds.), *The Black feminist reader* (pp. 24–56). Boston: Blackwell.

Morrison, T. (2008). *Toni Morrison: What moves at the margin: Selected nonfiction*. In C. C. Denard (Ed.). Jackson: University Press of Mississippi.

Narogin, M. (1995). *Us mob: History, culture, struggle: An introduction to Indigenous Australia*. Australia: Angus and Robertson.

Ogundipe-Leslie, M. (1985). Women in Nigeria. In D. L. Badejo (Ed.), *Women in Nigeria today* (pp. 119–131). London: Zed Books.

Ogunyemi, C. O. (1996). *Africa wo/man palava: The Nigerian novel by women*. Chicago: University of Chicago Press.

Okpewho, I., Davies, C. B., & Mazrui, A. A. (1999). *The African diaspora: African origins and New World identities*. Bloomington: Indiana University Press.

Olajubu, O. (2003). *Women in the Yoruba religious sphere*. New York: State University of New York Press.

Ologunde, A. (1982). The Yoruba language in education. In A. Afolayan (Ed.), *Yoruba Language and Literature* (pp. 277–290). Ife: University of Ife Press.

Olupona, J. K. (Ed.). (1991). *African traditional religions in contemporary society*. Minnesota: Paragon House.

Olopona, J. K. (Ed). (2000). *African spirituality: forms, meanings, and expressions*. New York: Crossroad.

Ong, A. (1999). *Flexible citizenship: The cultural logics of transnationality*. Durham, NC: Duke University Press.

Owomoyela, O. (1997). *Yoruba trickster tales*. Lincoln: University of Nebraska Press.

Oyewumi, O. (1997). *The invention of women: Making an African sense of Western gender discourses*. Minneapolis: University of Minnesota Press.

Oyewumi, O. (Ed.) (2003). *African women and feminism: Reflecting on the politics of sisterhood.* Trenton: Africa World Press.

Pieterse, J. N. (1992). *White on Black: Images of Africa and Blacks in Western popular culture*. New Haven: Yale University Press.

Pulis, J. W. (Ed.). (1999). *Religion, diaspora and cultural identity: A reader in the Anglophone Caribbean*. New York: Routledge.

Purcell, T. W. (1998). Indigenous knowledge and applied anthropology: Question of definition and direction. *Human Organization*, *57*(3), 258–272.

Raboteau, A. J. (2004). *Slave religion: The "invisible institution" in the antebellum South* (updated ed.). Oxford: Oxford University Press.

Rachewitz, B. (1964). *Black Eros: Sexual customs of Africa from prehistory to the present day*. London: Allen and Unwin.

Ransby, B. (2003). *Ella Baker and the Black freedom movement: A radical democratic vision*. Chapel Hill: University of North Carolina Press.

Razack, S., & Fellows, M. L. (1998, Spring). The race to innocence: Confronting hierarchical relations among women. *Journal of Gender, Race and Justice*, *1*(2), 335–352.

Richards, D. (1990). The implications of African-American spirituality. In M. Asante & A. S. Vandi (Eds.), *African culture: The rhythms of unity* (pp. 59–79). Lawrenceville, NJ: Africa World Press.

Richards, D., & Ani, M. (1994). *Let the circle be unbroken: The implications of African spirituality in the diaspora*. Ewing Township, NJ: Red Sea Press.

Rosenberg, D. G. (2000). Preface. In G. J. S. Dei, B. Hall, & D. Rosenberg (Eds.), *Indigenous knowledges in global contexts: Multiple readings of our world* (pp. xi–xvi). Toronto: University of Toronto Press.

Sedgwick, E. K. (1990). *Epistemology of the closet*. Berkeley: University of California Press.

Semali, L. M., & Kincheloe, J. L. (1999). *What is Indigenous knowledge? Voices from the academy*. New York: Falmer Press.

Smith, D. (1987). *The everyday world as problematic: A feminist sociology*. Toronto: University of Toronto Press.

Smith, L. T. (1999). *Decolonizing methodologies: Research and Indigenous peoples*. London: Zed Books.

Some, M. (1994). *Of water and the spirit: Ritual, magic, and initiation in the life of an African shaman*. New York: Penguin.

Some, M. (2003). "Gay African spiritual masters: The gatekeepers; An interview with Malidoma Some." Retrieved from www.malidoma.com

Some, S. (2003). *Falling out of grace: Meditations on loss, healing and wisdom*. El Sobrante, CA: North Bay Books.

Soyinka, W. (1959). *The lion and the jewel*. Alexandria: Alexander Street Press.

Soyinka, W. (1976). *Myth, literature and the African world*. Cambridge: Cambridge University Press.

Soyinka, W. (1981). *Ake: The years of childhood*. New York: Vintage.

Soyinka, W. (2012). *Of Africa*. New Haven: Yale University Press.

Spivak, G. C. (1990). *The postcolonial critic: Interviews, strategies, dialogues*. S. Harasym (Ed.). New York: Routledge.

Steady, F. C. (1987). African feminism: A worldwide perspective. In R. Terborg-Penn, S. Harley, & A. Benton (Eds.), *Women in Africa and the African diaspora: A reader* (pp. 3–24). Washington, DC: Howard University Press.

Stuckey, S. (1987). *Slave culture*. Oxford: Oxford University Press.

The Combahee River Collective Statement. (1983). In B. Smith (Ed.), *Home Girls: A Black feminist anthology*. New York: Kitchen Table - Women of Color Press Inc.

Thiong'O, N. W. (1986). *Decolonising the mind: The politics of language in African literature*. Nairobi: East African Education.

Thiong'O, N. W. (1993). *Moving the centre: The struggle for cultural freedoms*. London: James Currey.

Thompson, R. (1984). *Flash of the spirit. African & Afro-American art & philosophy*. New York: Vintage.

United Nations Educational, Scientific and Cultural Organization (UNESCO). (2014). Intangible cultural heritage: Ifa divination system. Retrieved from www.unesco.org/culture/ich/RL/00146

Vansina, J. (1965). *Oral tradition: A study in historical methodology*. London: Routledge.

Vansina, J. (1985). *Oral tradition as history*. Madison: University of Wisconsin Press.

Warner-Lewis, M. (1991). *Guinea's other suns: The African dynamic in Trinidad culture*. Dover: Majority Press.

Warner-Lewis, M. (1997). *Trinidad Yoruba: From mother tongue to memory*. Trinidad and Tobago: Press University of the West Indies.

Warner-Lewis, M. (2003). *Central Africa in the Caribbean: Transcending time, transforming cultures*. Trinidad and Tobago: University of West Indies Press.

White, A. M. (2007, Summer). All the men are fighting for freedom, all the women are mourning their men, but some of us carried guns: Fanon's psychological perspectives on war and African women combatants. *Signs: Journal of Women in Culture and Society*, *32*(4), 857–884.

Williams, A. D. (2008). *Inside the divine pattern: discovery of truth, wisdom and prophecy*. Calgary: Gemini 11 Inc.

Williams, P. J. (1997). Spirit-murdering the messenger: The discourse of finger pointing as the law's response to racism. In A. K. Wing (Ed.), *Critical race feminism: A reader* (pp. 229–236). New York: New York University Press.

Young, J. R. (2007). *Rituals of resistance: African Atlantic religion in kongo and the lowcountry South in the era of slavery*. Baton Rouge: Louisiana State University Press.

Youngblood, R. F., et al. (Eds.). (1995). *The new illustrated Bible dictionary*. Vancouver: Thomas Nelson.

BLACK STUDIES & critical thinking

ROCHELLE BROCK &
RICHARD GREGGORY JOHNSON III,
Executive Editors

Black Studies and Critical Thinking is an interdisciplinary series which examines the intellectual traditions of and cultural contributions made by people of African descent throughout the world. Whether it is in literature, art, music, science, or academics, these contributions are vast and far-reaching. As we work to stretch the boundaries of knowledge and understanding of issues critical to the Black experience, this series offers a unique opportunity to study the social, economic, and political forces that have shaped the historic experience of Black America, and that continue to determine our future. Black Studies and Critical Thinking is positioned at the forefront of research on the Black experience, and is the source for dynamic, innovative, and creative exploration of the most vital issues facing African Americans. The series invites contributions from all disciplines but is specially suited for cultural studies, anthropology, history, sociology, literature, art, and music.

Subjects of interest include (but are not limited to):

- EDUCATION
- SOCIOLOGY
- HISTORY
- MEDIA/COMMUNICATION
- RELIGION/THEOLOGY
- WOMEN'S STUDIES
- POLICY STUDIES
- ADVERTISING
- AFRICAN AMERICAN STUDIES
- POLITICAL SCIENCE
- LGBT STUDIES

For additional information about this series or for the submission of manuscripts, please contact Dr. Brock (Indiana University Northwest) at brock2@iun.edu or Dr. Johnson (University of San Francisco) at rgjohnsoniii@usfca.edu.

To order other books in this series, please contact our Customer Service Department:

(800) 770-LANG (within the U.S.)
(212) 647-7706 (outside the U.S.)
(212) 647-7707 FAX

Or browse online by series at www.peterlang.com.

www.ingramcontent.com/pod-product-compliance
Ingram Content Group UK Ltd.
Pitfield, Milton Keynes, MK11 3LW, UK
UKHW021839100926
8556IPUK00004B/436

9 781433 126093